Voluntourism and Multispecies Collaboration

Critical Green Engagements

Investigating the Green Economy and Its Alternatives

Jim Igoe, Molly Doane, Tracey Heatherington, Melissa Checker, José Martínez-Reyes, and Mary Mostafanezhad
SERIES EDITORS

Voluntourism *and* Multispecies Collaboration

Life, Death, and Conservation in the Mesoamerican Barrier Reef

Keri Vacanti Brondo

THE UNIVERSITY OF ARIZONA PRESS
TUCSON

The University of Arizona Press
www.uapress.arizona.edu

We respectfully acknowledge the University of Arizona is on the land and territories of Indigenous peoples. Today, Arizona is home to twenty-two federally recognized tribes, with Tucson being home to the O'odham and the Yaqui. Committed to diversity and inclusion, the University strives to build sustainable relationships with sovereign Native Nations and Indigenous communities through education offerings, partnerships, and community service.

© 2021 by The Arizona Board of Regents
All rights reserved. Published 2021

ISBN-13: 978-0-8165-4260-4 (hardcover)

Cover design by Leigh McDonald
Cover photo by Andrea Izaguirre
Typeset by Sara Thaxton in 10/14 Warnock Pro with Trade Gothic Next LT and Baskerville URW

Publication of this book is made possible in part by support from the Fine Arts, Humanities, and Social Science Grant program sponsored by the Division of Research and Innovation at the University of Memphis, and by the proceeds of a permanent endowment created with the assistance of a Challenge Grant from the National Endowment for the Humanities, a federal agency.

Library of Congress Cataloging-in-Publication Data
Names: Brondo, Keri Vacanti, author.
Title: Voluntourism and multispecies collaboration : life, death, and conservation in the Mesoamerican Barrier Reef / Keri Vacanti Brondo.
Other titles: Critical green engagements.
Description: Tucson : University of Arizona Press, 2021. | Series: Critical green engagements : investigating the green economy and its alternatives | Includes bibliographical references and index.
Identifiers: LCCN 2021004390 | ISBN 9780816542604 (hardcover)
Subjects: LCSH: Conservation of natural resources—Honduras—Utila Island. | Volunteer workers in conservation of natural resources—Honduras—Utila Island. | Ecotourism—Honduras—Utila Island. | Utila Island (Honduras)—Social conditions—21st century.
Classification: LCC QH77.H62 B76 2021 | DDC 333.72097283/15—dc23
LC record available at https://lccn.loc.gov/2021004390

Printed in the United States of America
♾ This paper meets the requirements of ANSI/NISO Z39.48-1992 (Permanence of Paper).

In memory of Miss Iris Hill, Mr. Cleveland Muñoz, and Mr. Shelby McNab.

Contents

	List of Illustrations	ix
	Preface and Acknowledgments	xi
	Introduction: Toward a Political Ecology of Multispecies Voluntourism	3
1.	"And the Sea Shall Hide Them": Utila's Cultural Landscape and Multispecies Entanglements	34
2.	"If You Come to Utila, You Can Do What You Want, and You're Never Gonna Leave"	70
3.	Conservation, Volunteering, and the Spectacle of Affective Labor	112
4.	The Political Ecology of Multispecies Conservation Voluntourism, and Limitations to "Becoming With"	148
	Conclusion: Life, Death, and Collaboration in Utila's Affect Economy	167
	Notes	*171*
	References	*181*
	Index	*193*

Illustrations

1. Preparing cotton ball scent traps. — 6
2. Live orchid bee, moments before its death. — 7
3. Preparing to measure and sex juvenile iguana. — 8
4. Reproduction of map of Utilian sites in Robinson Crusoe's journal. — 35
5. Camponado, 2020. Photo by Josely Turcios. — 36
6. Photo of Robinson Crusoe exhibit. — 37
7. Garden fencing surrounding turtle nesting site on Pumpkin Hill beach, December 2019. — 39
8. Absurd cover choice for map of Utila island published by magazine named after the colonial name for Pech, *Päyä: The Roatan Lifestyle Magazine*. — 77
9. Screenshot of Dr. John in his house posted in John DuPuis's online article "A Visit to Famous Dr. John in Utila," 2017. — 80
10. Screenshot from *If You Come to Utila*. — 83
11. Tradewinds homes with mangroves across the road. Prior to construction, mangroves reached the shoreline. — 102
12. Camponado lot under construction for new home. — 107
13. Framed photo of the 1990s turtle project, positioned in a central location in an islander's home in the Camponado. — 121

14. Exhumed turtles in 2016. Notice the two cameras taking photos as the BICA staff member exhumes the nest. My camera makes a third, and at least three other tourists were present with their cameras. 122
15. Screenshot of WSORC social media. The "whale shark" wetsuit makes its rounds among interns. 135
16. Extracting first boa constrictor from cage, preparing to remove ticks. 154
17. Removing ticks from boa constrictor by tweezers and fingernails. 154
18. Author and her husband departing for sweep netting. 157
19. Highlander toe collected and saved for genetic testing. 159
20. Iguana toe to be discarded. Cut for future tracking purposes. 159
21. Author performing iguana measurements. 160
22. Logging measurements. We always take a photo of the data next to the individual iguana. The bead tag is evident in the photo. 160
23. Mangrove propagules. 162
24. Journey through the water that hangs on author's office wall. 162

Preface and Acknowledgments

This first section is always the last section of anything I write. It is probably the same for most authors, as it is a moment of closure and reflection, a time to sit and reflect upon the relationships that made the work possible. The morning I wrote this, I woke up to a WhatsApp message from a dear friend from Honduras, whom I met in conservation circles in Utila. It said: "Good morning!!! I dreamt we were in Kentucky, on a trip jajaja . . . I fed you [a] half cooked chicken and you told me off . . . But in a friendly way because we were at the beach." I have been talking a lot about reciprocity with my graduate students this semester. As I finished this book, Lynn Bolles (2020) published a beautiful piece on "reciprocal arrangements" that chronicles a thirty-four-year friendship that began during Bolles's fieldwork in Jamaica. The piece reminds us of the myriad ways in which we are in reciprocal relations with the people we represent in our writing and that if we are lucky, reciprocal relations of mutual respect and care develop early and stay for the long term. That WhatsApp message captures perfectly the beautiful friendships that have grown out of the research. We were in Kentucky, close in my friend's dreamlike state to where I am physically present today (Memphis, Tennessee). We shared a meal, as friends do. I told him off, as friends can. Who knows why I told him off. Maybe it was about disagreements in conservation practices. But probably not. Maybe it was because the chicken was undercooked. But probably not. In the end, it did not matter. Because we were on the beach and because we were together. This friend was sending me a WhatsApp message from Tegucigalpa, where he was at the moment. He was dreaming of me in the southern United States, where I am at this moment. And we were brought together on a beach, likely on the island of Utila, a place we both love and care for deeply. This is a book about reciprocity, friendship, love, and care. These states comes in many different forms, from the interpersonal relationships so important to human sociality, to the love and care by humans for whale sharks, iguanas, and boas, to the symbiotic relationship found between mangroves and crabs.

Many life-forms exhibited care and kindness in making this book possible. These beings include the multiple species found within Utila's bountiful ecosystems that welcomed me, tolerated me, ran, swam, or slithered away from me. I cannot name all the islanders who shared their time with me individually, as doing so would breach my promises of confidentiality, but I hold the deepest gratitude and respect for these friends and families. I wish to thank all of the public figures in conservation and community work who shared their time and expertise over the years, including Andrea Albergoni, Dom Andradi-Brown, Edo Antúnez, María Arteaga, Ely Augustinus, Rossella Battaglia, Tom Brown, Luis Chévez, Steve Clayson, Suriel Dueñas, Steve Dunbar, Chris Howard, Andrea Izaguirre, Nilda Lino, Andrea Martínez, Daisy Maryon, Shelby McNab, Kate Meyer, Lori Muñoz, Pamela Ortega, Flavia Papini, Daniela Sansur, Josely Turcios, Junior Williams, and all of BICA's Reef Leaders. I have learned and continue to learn so much from each of you. Thank you for your patience with all my questions, and for my asking the same ones over and over again because I seem to have a permanent block for absorption of biological processes. Many of you have become very close friends and have welcomed my family members and me into your homes and lives during our travels. My family and I treasure these relationships born in Utila that now extend throughout the globe.

The research for this book spans two decades and has been supported by many institutional partners along the way. The 2002 data were supported by a Fulbright IIE, and Michigan State University's Department of Anthropology and Center for Latin American and Caribbean Studies. From 2011 forward, support for my work in Utila has come through a myriad of sources, including the University of Memphis's (UofM's) Department of Anthropology, the UofM's College of Arts and Sciences Travel Enrichment Grant Program, the UofM Engaged Scholarship Research Grant Program, the Center for Collaborative Conservation at Colorado State University, and the National Geographic Society. The receipt of a 2019 Professional Development Award from the UofM gave me the much-needed time to focus fully on writing this book, and I am grateful to the university for investing in my scholarship in this way. Portions of chapters 3 and 4 are derived, in part, from two articles published in the *Journal of Sustainable Tourism* in 2015 and 2018, available online: https://www.tandfonline.com/doi/full/10.1080/09669582.2015.1047377 and https://www.tandfonline.com/doi/full/10.1080/09669582.2018.1477784.

I am deeply indebted to the following scholars for this work, who provided important critique over the years on conference presentations and article drafts that eventually became part of this book or as members of the Critical Green Engagement editorial collective: Melissa Checker, Molly Doane, Andrea Freidus, Tracey Heatherington, Jim Igoe, Melissa Johnson, Suzanne Kent, José Martínez-Reyes, Mary Mostafanezhad, and Noel Salazar. This book draws on the work of William Davidson, who I recently learned lived nearby and whom I have since come to know personally: I am so grateful for your generosity and to be able to continue to learn from you. I also extend my thanks and gratitude toward the anonymous reviewers for this manuscript and associated journal articles. Your thoughtful feedback helped me to refine my arguments. It was a pleasure to return to work with the University of Arizona publishing team for a second time. Allyson Carter, Amanda Krause, Leigh McDonald, Abby Mogollón, and Sara Thaxton made the process run seamlessly.

Several graduate and undergraduate students worked with me over the years, each on unique projects that relate to themes in this book. Many of them worked on projects not featured in these pages, but their thinking and engagement with Utila contributed to my own thinking and provided me with a wonderful opportunity to stay in the headspace of such a special place. Chris Cosby, Cheri Foster, Ryan Kilfoil, Ivan Ortiz, Gretchen Pederson, Kat Robinson, Risako Sakai, Rachel Starks, Daryl Stephens, Josh Swiatek, Megan Warren, and Johnda Washington: in addition to your academic contributions and critical analyses, as a collective, you taught me the importance of maintaining lightness, encouragement, and presence. I have a lot of work to do in these areas, but you are all excellent role models for living a life of constructive engagement. Students in the following classes also participated in remote, collaborative research with island conservation organizations: fall 2015, Gender and Environment; spring 2016, Directed Readings on Conservation Volunteering; fall 2017, Cultural Perspectives on the Environment;

and, spring 2019, Culture, Conservation and Environmental Change. Arleen Hill cotaught the 2015 Gender and Environment course and generously dedicated her time and energy to thinking about the relationship between gender and collaborative conservation in Utila and beyond that semester. Chris Cosby played an integral role in the final steps of manuscript submission. With his eagle eye and patience, the last formatting pieces of the project came together swiftly. Moreover, I lucked out in that Chris took on a research project on conservation in Utila in 2020 and I benefitted intellectually from his regular discussions with me as he processed his own data. I, along with several conservationists in Honduras and beyond, also ended the year with a new friend.

There are a few key friends who deserve special recognition, friends who tracked this project's emergence in 2011 and supported it throughout the past decade. Sara Bridges and Kyle Simpson: each of you share a piece of Utila even without having ever been there. This comes from what you have each taught me about reciprocal arrangements, friendship, and care. Thank you for taking this journey with me and for being such wonderful friends. And then of course there is my longtime dear friend Suzanne Kent. We began our ethnographic journeys in Central America outside of Utila, but we did travel there socially a few times together in the early 2000s. Little did we know at the time that we would end up partnering on research there in so many unique ways. I hold so much respect and gratitude for you Suzanne and cannot thank you enough for walking with me over the years.

Utila has been a place held close in the hearts of all of my family. It was our "getaway" in the early 2000s when we lived on the north coast during my dissertation research. It was affordable, laid-back, and quiet. There was not yet twenty-four-hour electricity, and transportation was largely by foot and bicycle. The population was still fairly low in numbers and the community not yet saturated with tourists. Keegan "Kiki" Vacanti spent many months on the island as a baby, and he loved to watch the daily airplanes land, first on the shoreline by Bando Beach and later up on Pumpkin Hill after the new airport was constructed. It would be a long time before Amalie got to know the island—not until 2011, but she embraced and fell in love with it just the same, and from that date forward she was a regular visitor and fieldwork partner. Thank you all for getting muddy, for suffering through the sandflies and mosquitos, and for living in the treehouse. I am sorry I did not tell you I knew about the tarantulas. Kurt, Jamie, and Ursa: thank you for joining us on the island so many times. Iris, we hope you will spend time with us there someday. Ursa, I'm sorry that the beach came with sand. DJ and Keegan: while I know the cleaner ants were a little scary, I am glad you got to experience them, and I know that you are (now, in retrospect) also glad that you did. Amalie, my sweet empath, even if you end up choosing another path, I can already say you are an *incredible* anthropologist.

Voluntourism and Multispecies Collaboration

Introduction

Toward a Political Ecology of Multispecies Voluntourism

Utilla, 1904

The natural scenery in the island is common to the tropics—rich beyond description. A luxuriant vegetation greets the eye in almost every direction. Stately palms; lovely ferns in considerable variety; orchids, their bare long spires tipped with rich flowers of variegated hues; creepers of different sorts; and many beautiful wild flowers are scattered with nature's lavish profusion everywhere throughout the island.

Along the shore the soil is well adapted to cocoanut growing . . . But for the "cocoanut bug"—which is a large beetle of the Scarabaens family—the island would be a magnificent cocoanut grove from end to end. The planters planted again and again but the bugs killed as fast the trees grew. This beetle destroys hundreds, and some years, perhaps, thousands of the finest trees . . . Boring into the tree, the bug deposits its larvae and the fate of the tree is sealed. Whole cocoanut gardens in the interior of the island have been destroyed by this beetle, even before producing a crop . . . Several unsuccessful and somewhat spasmodic attempts have been made by the inhabitants to exterminate this enemy of the palm in our island.

The island furnishes excellent sea-bathing in some places. In good weather the water is clear as crystal, and the bottom of the sea may be clearly seen through the many fathoms of the transparent liquid. How delightful on a calm day to watch the fishes below,—many of them combining all the colors of the rainbow—darting hither and thither, and out through the rocks! Or it may be, perhaps, gaze admiringly and covetously at the branches of coral,—of every imaginable shape,—growing up from the bottom, but by the tireless polyp. (Rose 1904, 148–50)

Rose, who penned this rich description of Utila's bountiful landscapes in 1904, fashioned himself as someone for whom "book-making [was] not [his] calling." Yet he felt his fellow islanders shared an interest in preserving "the story of their beloved and picturesque islands," and so Mr. Rose took on this project of writing the first and only published nonfiction work documenting the island's history from the perspective of an early white settler inhabitant (preface, n.p.). Mr. Rose writes as someone who was truly taken by Utila's natural landscape, in awe of the ability of some species to proliferate and for others to counter human efforts to eradicate them. At the time his book was composed, Utila did not have any electricity, roads, or visitors. The population was almost 800, up from just 101 in the 1858 census (Rose 1904, 11). While Rose speculated that "the island may yet become a good resort for tourists from the North," during his time, there was not a single hotel or boardinghouse on the island. This situation would change drastically over the next century as Utila, Honduras, reached a global profile as the cheapest place in the world to get scuba certified.

Located approximately thirty kilometers from the Honduran mainland and on the edge of the Mesoamerican Barrier Reef, the second-largest reef system in the world, the Bay Islands consists of the main island of Utila and eleven small offshore Cays, a diver's paradise.[1] As dive tourism and associated infrastructure development have taken off to accommodate increased visitation to this small island, so too has its corollary: conservation, and now conservation voluntourism. Conservation voluntourism falls within Wearing's (2002, 240) early definition of volunteer tourism, which he described as travel focused on "aiding or alleviating the material poverty of some groups in society, the restoration of certain environments or research into aspects of society or environment." I use the term "conservation voluntourism" to describe those who pay host organizations to work on their own or others' species-specific research projects, to learn ecological monitoring methodologies, to feed and care for animals as part of captive breeding programs, to capture and tag endangered species, or to manage invasive species through culling.[2] In the case of Utila's conservation voluntourism industry, this form of humanitarian travel includes encounters with endangered iguanas, sea turtles, whale sharks, mangroves, and lionfish, to name a few.

A Typical "Volunteer" Day

On a hot summer morning in July 2016, my husband and two children join the "invert team" of a conservation research organization on their population study of orchid bees at the top of Pumpkin Hill on the Bay Island of Utila.[3] We set scent traps, squeezing a few drops of a distinct scent on three separate cotton balls that we then tie to tree branches in three separate locations, about fifteen feet between each (see figure 1). Then we sit down and wait. And wait. And wait. We pass the time listening to the team crack bee puns: "BEElieve me, we are waaayy more fun than the verts" (referring to their iguana-loving counterparts). "Yes, BEE-ing up here with us is *far* better than trekking through the stinky black mangroves." And we learn everything we can from them about orchid bees, especially curious about the shiny metallic blue *Euglossa*, found only in the Americas, and the ones we are hoping will land on our traps. None of us knew much of anything about orchid bees before this morning, but we learn that there are more than two hundred different known species of orchid bees and that new species are found every year. It was during our walk up the hillside that we learn we will only capture males; only males will fall for this trap: "females are always the smarter species," Amy ribs.[4]

The *Euglossines* got their common name "orchid bees" because of the behavior of males, who are constantly seeking the perfect smell, collecting scents to woo females into mating. Neotropical orchids have adapted to exploit this behavior by producing strong scents such as vanilla or eucalyptus, to ensure their pollination. Female orchid bees mate just once in their lifetime, so males will visit several different odor sources to build up an attractive collection of scents, storing them first in special organs behind their back legs, coating them in a fatty

Figure 1 Preparing cotton ball scent traps. Photo by author.

substance within those organs, and then spreading the scent to their front legs. From here, the bees continue to move the scent to their extra-large back tibiae, where the fatty scent mix is transferred through a small tube into the hollow interior of the tibias called the hind-tibial organ (HTO), and the scent is stored until the male is ready to mate. When he is ready, he will display himself through a performance where he hops and fans his wings, releasing his scent. If a female passerby likes what she smells, they will pair for reproduction.

There is a lot of talk about publishing opportunities; scientific studies of Utila's orchid bee population are virtually nonexistent. There is great potential to stake a career claim here. About forty-five minutes pass, and we all take turns monitoring the traps as others go on short walks to the side of the hill to stare out among the orchids and across the island's hardwood and mangrove forests; not a person or structure can be seen from this view. The affect expressed between the orchid bee and the orchid is what makes this view—and our experience of "becoming with" possible. Places and views such as these remind us of Donna Haraway's (2003, 244) observation that "if we appreciate the foolishness of human exceptionalism then we know that becoming is always becoming *with*, in a contact zone where the outcome, where who is in the world, is at stake." Who will persist in this landscape a century from now?

One hour and fifty-seven minutes have passed, and we will be taking our traps down in just three minutes, calling it a day, when—just in time!—an orchid bee lands on one of the cotton balls. Amy catches it by net and passes it around the members of my family so we can feel how he vibrates in our hand (figure 2). Males do not have stingers, we are reassured. This is the first time any of us have ever held a bee, and its forceful vibrations take us by surprise. Is this bee attempting to pollinate us as it does the orchid? Is it attempting to give us life or is it fearing for its own? We pass him around, each noticing the yellow pollen it leaves as it flaps its tiny wings, a sweet scent from the orchid. Moments later, Amy puts him in a tube and drops a bit of rubbing alcohol in, killing our new beautiful blue friend. I silently observe the irony of killing in conservation work. My husband gently probes about the numbers of bees on the island, and we are reassured that orchid bees are plentiful, and the science is critical if we are ever to learn anything about their population and diversity.

After a delicious lunch of pasta with an avocado-based sauce and game of Bananagrams back at the research station, with five researchers fifteen to twenty-five years my junior, my family leaves for town and I head out with the "vert team."[5] We walk for about thirty to forty minutes out to undeveloped mangrove-covered shoreline, where the team often encounters juvenile *Ctenosaura bakeri* (referred to locally as "swamper" or "wishiwilly"). I spend the first bit of my time trying—unsuccessfully—to locate and capture juveniles that are hiding among the groundcover. Fast forward to my discovery that Alice, the lead herpetologist, has found a nest on the shore and is busily excavating the babies, recording their measurements. She grants me the role of research assistant as she soothes the sixth of the seven juveniles the nest produces.

Figure 2 Live orchid bee, moments before its death. Photo by author.

"It's okay, I have you, this won't hurt," she coos, inserting a long probe into the pockets of the swamper's genitals to determine his sex (figure 3). "Another male!" I mark the sex in Alice's field notebook. Later we will cut off its left toe for genetic testing, and eventually Alice will know if the probing gave us an accurate read. Once this spiny-tailed iguana has been measured, weighed, sexed, and tagged, we release him onto a mangrove stump nearby. He doesn't move. "He's in shock. They do this a lot—just stay in the place we release them; I think he is stunned . . . [C]an you imagine if this was how you experienced your first moments of life outside of the nest? Welcome to the world, little guy. Poor thing." We—myself and two other women[6]—carry on recording data for all the juveniles we can find on the beachfront, while a team of three, one man and two women, conduct a transect survey deeper in the mangrove forest. This was the second of the two excursions that group made that day, ending with more than a dozen newly documented and sometimes tagged[7] swampers to add to the organization's dataset, and to the data pool that will be drawn upon by Alice and two others for their master's and doctoral theses. We also produced dozens of photos, of swampers, snakes, and interesting insects that we encountered along the way, passed around for volunteers to handle.

Figure 3 Preparing to measure and sex juvenile iguana. Photo by author.

Both groups that day, as most other days, were constituted by foreign researchers and volunteers. Even the full-time staff of this particular conservation organization had another country they called home, where they returned when Utila's tourism season came to a lull (October–December). Their extended presence, however, and interactions with the nonhuman species they encounter while present on the island, are transforming conservation and development narratives and protected area management in new, yet to be fully understood ways. Nearly twenty years prior to that day, our landlord—a descendant from one of the original Utilian families—greeted us outside the home we rented on her property holding a hammer, out to knock blue crabs on the head to take home and prepare them for a delicious dinner. Back then blue crabs were plentiful, found all over the main strip of town. Today harvesting blue crab is strictly regulated, and you won't find many locals inviting visitors to their homes to eat crab.

❖ ❖ ❖

The above scenes are connected in their expressions of affective labor in a "more-than-human" world (Braun 2004, 2006), where social life does not simply comprise human actors but instead is a complex assemblage of bodies and beings. Yet, the above accounts are human testimonies that attempt an understanding of a social world on a small island off the coast of Honduras that has experienced a long history of engagement with a variety of bodies (fish, crustaceans, iguanas, humans, and other life-forms), an island that has built its economy up around human relationships with each other, and with other beings. Utila, Honduras, the site of this ethnography, is a small but materially and socially complex place.

Bay Islander Ethnicity, Economy, and Nationality

Chapter 1 covers the islands' historical context in more depth. Here I provide an introduction to the diverse racial and ethnic groups that settled the island and briefly highlight how the area's changing economy has shaped human and more-than-human relations. The Bay Islands were first inhabited by the Indigenous Pech (Paya), who sustained their community through hunting, fishing, and small-scale cultivation. With Spanish contact in the 1500s, the Indigenous population was killed or captured and enslaved. The Bay Islands then became a focus of rivalry and armed conflict for more than two hundred years between England and Spain and served as primary sites for buccaneers and British pirates to hide their plunder and careen ships before raids to the mainland or Spanish ships (Stonich 1999).

The first settlers after the decimation of the Indigenous population came by way of the Cayman Islands and the United States in the 1830s. These first arrivals were white American and Anglo-Antillean peasants who were hungry for free, fertile land where they could eke out a living through subsistence agriculture. The first freed slaves to arrive in the Bay Islands also came from the Cayman Islands following the 1838 end to British slavery, settling first in Roatán. It was not until the mid-1850s that the first Afro-Antilleans were documented in censuses on Utila (Davidson, personal communication, October 22, 2020). Until then, Utila was settled almost entirely by Anglo-Antillean or Anglo-Americans who were fleeing emancipation in the United States, a fact that sets the tone for ensuing race relations on the islands.

A second group of English-speaking islanders of greater African descent arrived in the Bay Islands in the early decades of the twentieth century. This group comprised migrant workers from Jamaica and other Caribbean islands, including Belize, who were working for the banana companies on the north coast (Graham 2010). In addition to settlers from the Caymans and other Caribbean islands, there are a number of Afro-Indigenous Garifuna families and Honduran mestizos on Utila and its Cays; both groups have also lived in the Bay Islands for several generations now.[8] The Garifuna are descendants of an intermixture of marooned African slaves and native Amerindians (Carib and Arawakan), emerging as a unique racial and cultural group in the mid-seventeenth century on the island of St. Vincent. They were exiled from the island of St. Vincent to the Bay Island of Roatán in 1797, and from there they settled in coastal communities all along the north coast and islands.

Distinction based on "color" has long been a primary way in which Bay Islanders self-identify. When Afro-Antilleans arrived, they settled in distinct neighborhoods and these two groups lived both socially and spatially segregated for multiple generations. Phenotypically lighter-skinned Cayman islanders with African ancestry who migrated to the islands generally deidentified with their African origins, a practice that is common throughout the Caribbean and continues today in the Bay Islands. In his 1961 account of the ethnic heterogeneity in French Harbour, Roatán, David Evans (cited in Stonich 1999, 51) found that while most of the self-identified "whites" in his study knew of their African ancestry and would have been classified as "colored" or "Creole" (i.e., descendent of African and Anglo intermixture that extends for many generations), people went out of their way to disassociate with black islanders through spatial and social segregation. Just as in Roatán, segregation marked Utila's settler foundation. And, in its earliest years, Utila—unlike Roatán and Guanaja—was an almost exclusively white/Anglo-Antillean enclave. Even when black islanders and Garifuna settled on Utila, social and physical segregation kept communities apart, with separate churches, dances, and neighborhoods. In his 1970s fieldwork, anthropologist David Lord (1975, 107) documented a white Utilian saying, "This is Little Rock, Arkansas," implying that prejudice and segregation were part of the social fabric of Utila.

Culturally, white and black English-speaking Bay Islanders and Garifuna distinguish their collective identity from the Spanish-speaking mainland. While all ethnicities regularly partner with and marry Honduran mestizos, and mixed-descent families are commonplace, older islanders maintain a cultural affection for Britain and the United States and prefer to identify as "English" as opposed to mainland "Spaniards." Settlers from the mainland are segregated by class, and many of the more affluent mestizos have become "white" through marriage and the accumulation of property, matching the *blanquimiento* process of "whitening" through race mixture that devalues blackness (Wade 1997). Poorer mestizos remain at the bottom of the social hierarchy, described in more detail in chapter 1. Because of the complicated history and social hierarchy, and on the recommendation of an early manuscript reviewer, I identify the racial/ethnic background and positions of my interlocutors throughout the text.

While Spanish is the official language of Honduras, English had been present in the Western Caribbean since well before 1600 and was the primary language spoken by the first permanent settlers in the Bay Islands. The variation of English spoken on the islands is known as "Bay Islands English" (BIE). BIE is not a Creole per se, but rather a regional variety of English that has been influenced by contact with creolized English (Holm 1983, cited in Graham 2010). BIE can be understood as a spectrum of spoken varieties across the islands that have at their foundation a selection of English dialectal features from the Cayman Islands. These varieties are influenced by the "semi-Creole" of the Cayman Islands, with direct Creole inputs from Belize and Jamaica (Graham 2010, 100). According to Ross Graham (2010), as the minority group, Creole speakers in the Caribbean would have adopted their speech to the majority, which meant early settlers of the islands spoke English closer to "Standard English" but with a distinctive accent similar to Caymanian. The distinct accent of Utilian BIE speakers was originally contained in quotes I present from my island interlocutors. Not all islanders quoted in this book, however, speak with a strong BIE accent, especially if they spent a significant portion of their school and work lives offshore or in the United States. Out of caution and concern that I might misrepresent the phonology of the BIE dialect, and due to the variation among speakers, I have chosen to present islander transcriptions in Standard English.

English persisted as the preferred—and most commonly heard—language on the islands for most of their settled history, and for generations many islanders were monolingual English speakers. Today, only a very small number of "Old Heads" (i.e., descendants of one of the settler families) would claim to be English-only speakers (even though most of this small number can also communicate effectively in Spanish). Whereas in the early 1990s, Susan Stonich (1999, 121) found that only half of the Bay Island households she surveyed had parents who could speak Spanish, by 2000 84 percent of islanders spoke both English and Spanish. At the same time, only 33 percent of Spanish-speaking households could also speak English (Stonich 1999). Now, twenty years later, the rapid hispanicization of the islands has resulted in even fewer bilingual households, with Spanish-only speakers outnumbering bilingual islanders by leaps and bounds.

Not only did early islanders distinguish themselves culturally and linguistically from mainland Honduras, but they also preferred to orient their economy outward, with the majority making a living without having to work on the mainland. The first settlers on Utila constructed their homes on the Cays and developed small-scale agricultural production on the main island. The early agricultural economy centered on the cultivation of coconuts, bananas, plantains, mangoes, and other products for export to the United States (Rose 1904). New Orleans quickly became "the U.S. capital of Utila," or "Little Utila," as some locals say. With the growing trade between Utila, Belize, and the United States, a significant part of the Cayan population moved to the larger island so they could extend their plantations, and Utila enjoyed an agricultural boom between the late 1800s and early 1900s (Lord 1975). Agriculture declined with the rise of the United Fruit and Standard Fruit Companies, each planting large farms along Honduras' north coast and outcompeting production on the Bay Islands.

The blow to the agricultural economy coincided with a loss of identity as British subjects. Utilians long identified with the British Empire, and in 1849 the Bay Islands issued a petition to fall under British protection. Despite the infringement of the Monroe Doctrine and Clayton-Bulwer Treaty, which states that the United States and Great Britain may not occupy, fortify, or colonize any part of Central America, the British declared the islands a colony, and the Bay Islands Colony lasted from 1852 to 1858. Throughout this time, islanders enjoyed many benefits of this status, including the receipt of British land grants to properties they claimed and avoidance of duty taxes at Honduran ports.

The establishment of the British colony was brought to the attention of the United States, and eventually under a great deal of pressure, England agreed to surrender the islands to Honduras in 1860. At the time of its surrender, Utila's population was approximately one hundred (Rose 1904, 11). The transition to Honduras did not significantly impact early settlers—at least not at first—as they continued living under English Common Law for many decades. Utilians were permitted to retain all rights to the land they claimed as British citizens, were granted religious freedom, and retained positions of power within municipal offices.

While the Bay Islands were formed into the Department of the Republic of Honduras in 1872, and thus subject to the republic's laws and governance, islanders assumed they would continue to live outside of Honduran law in perpetuity. They did live free from enforcement of Honduran law until the turn of the century, when in 1902 Captain Cooper-Key was sent on the HMS *Pscyhe* "to disabuse Bay Islanders of their uniform pretenses to British citizenship" (Lord 1975, 41). Rose (1904, 35–36) documents the reaction of islanders once the transition from English Common Law to Honduran governance was complete:

> The (ensuing) change of laws gave a crippling blow, for some time, to the industries in the islands and to the hopes of the people. There was general discontent chiefly on account of the high import duties imposed under the new laws. And this discontent was perhaps excusable, because the people had always been accustomed to a very low tariff. Many of the people began to think seriously of leaving the islands, and some did so; but the majority loved too well the land which had been their home for many years, so they remained.

Most islanders still speak fondly of their European heritage and their protection by the British Crown; many self-identify with these roots to differentiate themselves from mainland Hondureños.

Around the time Utilians became Hondurans, they also moved more squarely into a new market. With trade in bananas disappearing alongside the growth of the U.S. fruit companies, islanders turned to the harvest and production of coconut oil. Anthropologist Lord (1975) describes the difficulty of the decade between 1929 and 1939:

> With no market for agricultural commodities, no wage labor available since job specializations had never developed (and with too small a population to support specialists in any event), and with shipping defunct, householders were reduced to rendering out coconut oil in order to survive. This was an arduous way of making a living. Two hundred select coconuts (four inches in diameter or larger) were husked at a time, shells chipped off, the meat grated and mixed with water. After standing overnight the coconut "milk" would be skimmed off (oil that had risen to the top) and this would then be boiled down to the final coconut oil: five gallons in all. The tin of oil was then taken to a local merchant who paid for it by giving the maker fifty cents (U.S.) worth of goods from his store!

Lord's description of the arduous labor of coconut oil production stands in sharp contrast to the fond memories Utilians have of the coconut groves and the oil they produced. Older generations reminisce about the difficult labor but sense of freedom that came along with the slow pace of life associated with small island culture. Nostalgia runs deep in Utila.

The agricultural economy declined in the 1940s, and merchant marine and other seafaring work became primary occupations for Utilian men (Currin 2002). The Second World War brought opportunities for Bay Island men to work on U.S. navy supply ships, and this work transitioned into work for American and Norwegian merchant shipping lines, oil tankers, and cruise ships (Graham 2010; Stonich 1999). Offshore employment provided the financial impetus for the development of commercial shrimp and lobster industries from the late 1950s onward (Graham 2010). Despite declining fisheries, fishing for subsistence and sale continued as an important source of income for local families. According to the latest census (from 2020), there were a total of 98 fishermen on the island: 35 on Utila and 63 in the Cays, where fishing continues to be the predominant activity (BICA 2019).

In the 1980s, the tourism boom began, leading to a population swell. The residential population grew from roughly fifteen hundred to more than seventy-five hundred. Today, island residents include not only the English-speaking Anglo and Afro-Caribbeans who originated from the Cayman Islands and Garifuna and from mestizo families but also increasing numbers of foreign "lifestyle migrants" and heightened numbers of poor mestizo mainlanders who migrated to the island in search of employment in the growing tourism economy (Benson and O'Reilly 2009). The island's colonial legacy has much to do with contemporary race relations on the island, with divisiveness both between Anglo-Antillean (white) and Afro-Antillean (black) settlers, and between those who trace their roots back to the British Crown and Cayman Islands, the foreign expatriate community who launched the scuba dive industry in the 1980s, and those who have migrated over from mainland Honduras to work within tourism.

Protecting an Uninhabited Space

Utila is often portrayed as being "largely uninhabited"; that is because those doing the describing think about habitation as uniquely human. Utila is, in fact, full of life. Seventy percent of this 17.37-square-mile island comprises mangroves and associated wetlands, home to several endangered and endemic species. Utila has four species of mangrove trees: white mangrove

(*Laguncularia racemosa*), red mangrove (*Rhizophora mangle*), black mangrove (*Avicennia germinans*), and buttonwood mangrove (*Conocarpus erectus*). The red and black mangroves are most common and are habitat for an endemic spiny-tailed iguana (*Ctenosaura bakeri*), or "swamper." *Ctensosuara bakeri* is listed on the International Union for the Conservation of Nature's (IUCN) Red List as critically endangered for threats to population due to "habitat loss and fragmentation associated with development for tourism [and] decreasing quality of habitat from introduced invasive vegetation and degradation of nesting habitat due to local and oceanic pollution," as well as local consumption and sale of eggs and meat (IUCN 2018; qtd. in Maryon et al. 2018, n.p.).

Utila's mangroves are also home to other iguana species, including the nonendangered *Iguana iguana* and the *Ctenosaura similis*, with the latter easily confused with the swamper by nonspecialists. In addition to the swamper, conservationists, biologists, and ecologists are also keenly interested in the endangered whale shark population (*Rhincodon typus*, locally referred to as "Old Tom"), the endangered green (*Chelonia mydas*) and loggerhead (*Caretta caretta*) sea turtles, and the critically endangered hawksbill sea turtle (*Eretmochelys imbricate*); the loggerhead and the hawksbill nest on Utila's beaches. The Mesoamerican Barrier Reef system is home to more than 500 fish species, 350 mollusk species, and 65 species of stony coral, and Utila hosts one of the world's only year-round whale shark populations.

The arrival of the lionfish (*Pterois volitans*) to the Caribbean has significantly altered coral cover and reduced native fish populations. Native to the Indo-Pacific, lionfish proliferate at an extremely rapid rate and consume native fish species. With reproductive maturity under one year and high fecundity (female lionfish can produce floating, unpalatable egg masses of twenty thousand to thirty thousand eggs every four days all year round), they have spread rapidly and are responsible for the reduction of important native fish and crustacean species (Andradi-Brown and Vermeij et al. 2017).

With the growth of tourism, Utila's impressive mix of ecosystems—including mangroves, beaches, and coral reef ecosystems—has come under intense development pressure. Environmental concerns on the island are numerous, including mangrove and coral reef destruction, overfishing, illegal capture, consumption and sales of endangered and endemic species (including sea turtle and iguana), and inadequate solid waste disposal and wastewater treatment systems. The island also has limited freshwater.

There are five locally active conservation organizations that work to mitigate ecosystem degradation and promote opportunities to engage in the science of conservation. Three are NGOs: (1) Bay Islands Conservation Association (BICA); (2) Fundación Islas de la Bahía (FIB), which runs the operations for the Iguana Research and Breeding Station; and (3) Kanahau Conservation Research Facility (or Kanahau for short). Two private conservation businesses also host conservation volunteers on the island: (1) Whale Shark and Oceanic Research Centre (WSORC); and (2) Operation Wallacea (Opwall), a UK-based company with sites across the globe. Kanahau follows a similar model to Opwall but is based solely on the island of Utila. It was founded by past FIB employees and staffed by former Opwall volunteer scientists. These organizations focus on recruiting volunteers from university settings who are interested in gaining field-based experience in ecological or environmental subjects. Student volunteers are recruited both nationally and internationally, with clear divisions between organizations. Opwall, for instance, only promotes their programs within universities, and more than 95 percent of their volunteers are undergraduate or master's students working on research theses. Demographically, volunteers and organizational leadership typically match quite closely, with BICA and FIB constituted largely by Hondurans from universities based in the capital city of Tegucigalpa or coastal city of La Ceiba, and Kanahau and WSORC reflecting a mix of students and staff from the UK and United States (Kanahau currently has a strong Honduran staff presence, and WSORC's new community outreach coordinator is Honduran). Coral Reef Alliance, a global nongovernmental organization (NGO) focused on community-based conservation of coral reef regions, is also represented on the island, with a part-time project manager based on Utila. All of the various organizations work cooperatively with each other and with the municipality's environmental office, Unidad Municipal Ambiental (UMA).

The first and longest-standing organization is BICA, which was founded in 1990 on Roatán by influential island residents concerned with conservation on the Bay Islands. Chapters were then established on the other two Bay Islands, Guanaja and Utila. The origin of BICA-Roatán aligned with Honduras' push to aggressively grow its tourism economy in the early 1990s through the creation of tourism zones. Conservation and tourism development became intertwined from then on for the Bay Islands, and conservation organizations played central roles in promoting tourism. In her study of conservation and tourism development on the islands in the 1990s, Stonich

(1999, 157) found that "the vast majority of BICA programs are aimed at maintaining an environment conducive to the promotion of tourism." Conservation voluntourism is an extension of this approach.

Supporting the entanglement of conservation and development, Honduras established a number of protected areas based on the IUCN classification system, as well as passed legislation to protect key species from overexploitation. The Bay Islands waters (extending twelve nautical miles around the coasts of the islands Guanaja, Roatán, and Utila) were designated as a national park in 1997 under the Executive Agreement 005–97 of the Honduran government. In 2009, Executive Agreement 142–2009 defined zoned categories for management, including two marine zones where fishing is limited to line fishing only, and one wildlife refuge. In 2010, the Bay Islands were recognized by the Honduran Congress as a national marine park, and three zones of special protection were created: the marine areas of Turtle Harbour–Rock Harbour and Raggedy Cay–Southwest Cay, and the terrestrial wildlife refuge of Turtle Harbour. UMA, BICA, and FIB serve as comanagers of the protected areas.

It is illegal to hunt Utila's *Ctenosaura bakeri* and to harvest many marine species, including sea turtles and their eggs, conch, and juvenile lobster; spearfishing is banned from all areas of the national park. Mangrove clearance is also prohibited; however, limited funding for monitoring and enforcement combined with a high level of corruption has meant that illegal logging of mangroves is common (Canty 2007; Harborne, Afzal, and Andrews 2001). Not all clearance is illegal either; certain development projects have been granted permits for mangrove clearance and canal dredging, including controversial projects on the south end of the island such as Oyster Bed Lagoon project (http://oysterbedlagoon.com) and Coral Beach Village, a three-hundred-acre major development project with home sites, a restaurant, marina, and hotel (https://coralbeachvillage.com/). In 2020, the latter was up for sale with a price tag of $5.95 million.

Limited government resources for conservation means that local organizations are understaffed and frequently unable to conduct ecological studies on their own. Financially strapped, they often spend the bulk of their human resources applying for grants and hosting scientists and conservation voluntourists from other parts of the globe. The latter are a central revenue source keeping local organizations afloat, and all of them are all working to enhance their volunteer programs in ways that will lead to financial sustainability. Both FIB and Kanahau have active terrestrial programs; the largest for each focuses on the *Ctenosaura bakeri*. BICA and WSORC are more marine based in their work, with one of BICA's largest efforts surrounding turtle monitoring, and WSORC focusing on whale shark tracking. BICA and WSORC work together with Opwall on lionfish ecology. All organizations have a variety of other ongoing marine and terrestrial research programs as well. The conservation scientists that come to advance ecological research become critical brokers to enabling affective encounters for general conservation voluntourists with other species.

Political Ecology, Tourism, and New Materialism

This book is grounded in a theoretical frame known as political ecology. Political ecology is an approach to inquiry that connects ecological change to political and social relations. Applying a political economy perspective to reveal who wins and who loses from environmental changes that stem from dominant development models, political ecologists are concerned with unveiling who bears the cost of environmental change and what that cost looks like.[9] Political ecologists attempt to capture and narrate the various claims being made to nature, especially the impact of capitalist development on the environment, the social and political implications of conservation and environmental management, the state of nature, or how new natures (new landscapes, ecosystems, species) are produced (Peet, Robbins, and Watts 2011, 24).

In advancing a political ecology of tourism, Mary Mostafanezhad et al. (2017) highlight three core elements of political ecology analyses. First, the work of political ecologists is *historically situated*. For scholars working on issues related to tourism, this implies close attention to coloniality and modernity, thus challenging a unilinear explanation for environmental change in a given place and calling for the writing of alternative histories and attention to local resistance to tourism development initiatives. A second feature of political ecology is that such studies are typically *place-based*. These works frequently employ an ethnographic approach and rely on fieldwork with local community members to present a more nuanced understanding of the ways in which meanings are constructed, paying close attention to *whose* meanings are heard and validated. In tourism contexts this means a focus on the ways in which gender, ethnicity, race, class, sexuality, disability, and other social positions intersect with the costs and benefits of tourism projects, and how the environmental impacts of tourism development are differentially felt depending on one's social position. A final feature of a political ecology analysis is the attention to *multiple scales*. Political ecologists focus on the connections between local communities and local decisions, state and regional power, and the global political economy (Mostafanezhad et al. 2017, 6–8; Robbins 2012).

In this book, I seek to advance a political ecology of multispecies voluntourism by confronting head-on a contemporary conundrum in the twenty-first-century green economy: that the continuously expanding global tourism industry be positioned to mediate the capitalist contradictions to which it is also a central contributor (Igoe 2017, 309; Mostafanezhad et al. 2017, 1). Simply put, conservation voluntourism has arisen as a capitalist endeavor to solve problems created by capitalism itself. Impelled to "save the environment"—or to see it before it vanishes—increasing numbers of tourists and researchers travel to fragile ecosystems such as those of Utila, Honduras, and here they enter into a variety of multispecies relationships constituted by affect. Following Bruno Latour's (2004) understanding of affect as a dynamic process occurring at the interface of beings, affect herein is not treated as an emotion, residing with a person or subject, but rather as something produced *between* humans and humans, humans and animals, or even between nonhuman species such as mangroves and corals, or iguanas and mangroves. Thus, I contend that the field of conservation voluntourism is both generated by and produces what I am calling an "affect economy," that is, an economy based on the exchange or trade in the relational or "becoming with" (Haraway 2008, 2016) and the production and consumption of emotions and feelings that emerge through close encounters with another species. This reading of affect is not personal but rather is transpersonal, drawing on many bodies. It exists beyond cognition and therefore beyond humanness (Pile 2010, 8). Through examples of whale shark, lionfish, iguana, turtle, and other encounters, I show how conservation organizations operate as affect generators, enabling the privilege of engaging in multispecies encounters, to be affected and to produce affect. Conservation voluntourists purchase these experiences, which rest on a sanitized and depoliticized suffering subject that circulates through spectacle and exchange the cultural capital they accumulate through affective encounters with other species for a new social status.

While many political ecologists are focusing on the relationship between human livelihoods, conservation, and development in tourism contexts,[10] new materialists in political ecology are working to comprehend the more-than-human aspects of the Anthropocene (e.g., Barad 2007; Haraway 2016; Kohn 2013; Tsing 2015). In assuming "a political ecology of multispecies voluntourism," this book seeks to theorize the emerging green economy of multispecies voluntourism that is predicated on both the exploitation and conservation of specific place-based species. The core questions confronted in this book are: How are human socialities made through interactions with other species? What lives and dies in Utila's affect economy? Why are some species killable? Who gets to decide?

There are two important currents in the argument presented in this book. First, I suggest that conservation voluntourism, a human product, is closely enmeshed in coloniality and capitalist frames. Capitalism is driven by accumulation by dispossession, or the taking of a people's land and labor to create profit. Capitalism is constantly in need of "new sites for sinking, or investing, surplus" (West 2016, 19). One way to find new sites for capital accumulation is through "rhetorical dispossession," or the use of language to displace and dispossess others of their land, labor, or resources; the language of the "frontier" or discourses of "discovery" are prime examples of rhetorical dispossession (West 2016). Rosa Luxemburg (2003) argued that capitalism abroad can behave in ways that it would not, and could not, at home. Drawing on Luxemburg's observation, Paige West (2016, 18) points out that capitalism can behave violently, killing anything it chooses, in far-off places where discourses of discovery are at play. Conservation voluntourism is built upon such rhetorical narratives, with the work of volunteers centered on the discovery and documentation of new species and their population diversity. The rhetoric that surrounds conservation also functions to dispossess those who live or had historically lived with and managed interactions with vulnerable species or nonhuman species that are otherwise interesting to humans. Attending to this form of displacement is a way of locating coloniality in the multispecies conservation landscape, as specific hierarchical orders are imposed that value particular forms of knowledge and ways of relating to nature over others. Multispecies voluntourism rests on the creation of value and a new "sink" for capital to enable some to accumulate resources and opportunities while others are dispossessed of their place within the local landscape and relationships with other species.

A second thread builds on the momentum of new materialists to analyze the growing conservation voluntourism industry through an appreciation for the role of other-than-humans in the Anthropocene.[11] New materialist inquiry pushes beyond a narrow and anthropocentric understanding of the human condition, recasting the dualistic subject/object, nature/culture, mind/body, mental/material divisions into a more-than-human ontology that does not prioritize the agency of any entities over others. It extends the ongoing deconstruction of the Cartesian human/nature divide with a shift toward thinking of materialities not as passive "things" but as active, entangled, self-organizing, and vital (Schulz 2017). The world, through this frame, is constituted by entanglements of many entities, organic and inorganic, unstable assemblages that are constantly in a state of emerging.

When we attend to the agency of other beings in writing about cultural encounters, we begin to see how diverse beings coinhabit and collaborate for survival in landscapes constituted by capitalistic destruction. In recognizing animal autonomy, we shift our attention from thinking about nonhuman animals as resources or banks of natural capital and instead consider them as "beings with their own familial, social, and ecological networks, their own lookouts, agendas, and needs" (Collard, Dempsey, and Sundberg 2015, 328). It is in this spirit that I wrestle with how different ways of "worlding" (Blaser 2013; Haraway 2016) in Utila may be in conflict and may produce inequalities among humans and other species, but also how these ways of worlding are not solely directed by human action. The notion of "worlding" emerges from a theoretical area known as political ontology, an approach coined by Mario Blaser (2009, 2013). If ontology is the philosophical consideration of the nature of being, political ontology is concerned with attending to power, conflict, struggle, resistance, sustainment, and imagination in world making. Arturo Escobar (2017, 66) describes the focus of political ontology in two ways: "On the one hand PO [political ontology] refers to power-laden practices involved in bringing into being a particular world or ontology; on the other hand, it refers to a field of study that focuses on the interrelations among worlds, including the conflicts that ensue as different ontologies strive to sustain their own existence in their interaction with other worlds."

What political ontology does for political ecology is provide an analytical approach that moves beyond the deconstruction of dualisms associated with modernity and its "truths" and toward reconnection and relationality.

These normative divides (i.e., human/nature, human/nonhuman, mind/body, subject/object, reason/emotion, living/inanimate, organic/inorganic) are part of a universalizing ontology that marks the dominant form of modernity associated with what John Law (2015) has termed a "one-world world." The one-world world is a way of worlding that has become so dominant that it is difficult to think and see beyond. Here is where political ontology extends a helping hand: such an approach makes a concerted effort to reconnect the divisions associated with modernity, to view the relationships between—and entanglements of—humans and nonhumans, nature and culture, and so forth. It makes visible the effects of a one-world world by uncovering the outcomes of accumulation by dispossession and attending to relational ontologies and other ways of worlding. This is an approach that scrutinizes human-centered assemblages, reads coloniality in the landscape, and thinks in possibilities (beyond the one-world world), even when the landscape assemblage under study is partially connected to the one-world world, such as is the case in this book. Thus, in this book, I borrow from the work of Ana Deumert (2019) and Laura Ogden (2011), who each approached landscape assemblages by thinking rhizomatically, understanding them as continuously emerging, transforming, and mobile in material reality and meaning. I do so such that we can begin to imagine alternative worldings in Utila, beyond the tourist-spectacle-dominated world attached to the volunteer tourism industry.

To summarize, a critical political ecology informed by new materialist approaches within multispecies ethnography sees coastal tourism settings, such as the one in focus in this book, as points of convergence where marine, terrestrial, and human systems come into contact. Emerging from this frame, this book confronts the role conservation research tourists play in local protected area landscapes, asking, What transformations in protected area management are emerging from multispecies encounters and inequalities in relation to caring? What alterations emerge as nonhuman species interact with each other and confront or deny the boundaries and interactions imposed through human decision-making? Who lives and who dies in conservation areas?

Voluntourism as Affective Labor

With wildlife populations and biodiversity riches threatened the world over, new and innovative methods of addressing these threats are necessary—and none, we are told, are newer and more innovative than those drawing and/or relying on "the market." (Fletcher, Dressler, and Büscher 2014, 4)

The edited volume *Nature*™ *Inc.* of the University of Arizona's Critical Green Engagement series features several examples how global conservation has been significantly refashioned by neoliberal capitalism, for example, via increased financialization and global circulation of "natural capital" through new markets associated with carbon trade, species banking, and payment for ecosystem services (Fletcher, Dressler, and Büscher 2014, 3–4). Of the many studies of the negative consequences of commercialized conservation featured in that volume and elsewhere,[12] I would add conservation research voluntourism to the list of market approaches. Thus, conservation voluntourism is a form of neoliberal conservation, an argument I have detailed elsewhere (Brondo 2013, 2015). For readers new to the phrase, neoliberal conservation refers to the decentralization of environmental governance, and new forms of commodification and commercialization of nature that emerge in order to fund conservation efforts (Igoe and Brockington 2007). It is, as Büscher (2012, 29, qtd. in Fletcher, Dressler, and Büscher 2014, 14) writes, "the paradoxical idea that capitalist markets are the answer to their own ecological contradictions."

A significant body of literature now exists to illustrate the negative on-the-ground effects of neoliberal approaches to environmental governance, including the physical displacement of local peoples and the obstruction of Indigenous knowledge.[13] These conditions have been particularly acute in tourism zones and protected areas. In such areas, the negative impacts of neoliberal environmental governance are accompanied by "conservation successes," including the expansion of specific wildlife populations and increased foreign revenue (Sullivan, Igoe, and Büscher 2013). But, as Sian Sullivan, Jim Igoe, and Bram Büscher (2013, 15) note, paradoxically, such "success" comes with growing environmental concerns due to rising emissions and climate change resulting from the pressures of heightened tourism. In the end, the paradox of neoliberal environmental governance is that it obscures and exacerbates both the structures and causes of environmental degradation and the micropolitics of resource use and conservation.

The descriptor "Nature™ Inc." is used to represent a "truly new global conservation frontier." The "Inc." (or, "incorporated") in the phrase is meant to signify the fact that within neoliberal conservation, nature stands apart from society and the economy as a distinct "corpus" or "entity" (Fletcher, Dressler, and Büscher 2014, 10). Indeed, this is how "nature" is presented in social media imagery used to promote conservation voluntourism in Utila. In using the ™ trademark denotation, the authors are emphasizing that the "nature" in question is seen as something that must be "protected" through legalized and institutionalized governance systems, justifying protected area management plans, and legislation. Finally, "nature" in their construct should be read as an entanglement of humans and nonhumans "within complex 'socionatures,'" emphasizing the agency of nonhumans as "actants in such networks rather than as passive objects of human manipulation" (Fletcher, Dressler, and Büscher 2014, 10). From this description, Honduras' affect economy—one based on multispecies voluntourism—can be read as a form of Nature™ Inc.

What kind of labor is valued in the twenty-first-century green economy, under Nature™ Inc.? Unlike in the past, when labor resulted in material outcomes, twenty-first-century capitalism is defined by immaterial labor of two sorts. The first form is labor associated with the analytical services of problem solving and creativity associated with knowledge-based jobs (Hardt 1999, 95; Reich 1991). Second, and that which is in focus here, is the "affective labor of human contact and interaction" (Hardt 1999, 95). "Affective labor," then, is work meant to produce emotional experiences in people. This form of care work focuses on "moments of human interaction and communication" (95–96). While the labor involved is affective and corporeal, it is *immaterial* in that it does not result in a tangible product. The product is a feeling, a sense, an essence.

The observation that affective labor produces value is not new; generations of feminists writing about care work have made this argument (Skoggard and Waterston 2015). However, what is new under twenty-first-century capitalism is the degree to which this form of immaterial labor has come to directly produce capital and spread so broadly, finding itself now in *the* dominant position within the global capitalist economy (Hardt 1999, 97). Extending Michael Hardt's analysis into the contemporary context of Nature™ Inc., I posit that such affective production can also be found within the global tourism industry. For my purposes, I am concerned specifically with affective production within the complex "socionatures" associated with conservation voluntourism, locatable as affect exchanges between human and nonhuman species. These affects—forms of immaterial labor—then take on an exchange value in the global market.[14]

"Affect" features prominently in the current phase of global capitalism, a period of time in which the majority of jobs are those that are highly mobile and require a flexible skill set. Many of them focus on caring. From health services to fast-food workers to the entertainment industry, this new economic landscape is dominated by jobs that rely on knowledge, communication, human interaction, and *affect* (Hardt 1999). Volunteer tourism provides people with the opportunities to cultivate these qualities, making them attractive in this emerging workforce (Butcher 2003; Vrasti 2012).

There is a wide literature on motivations for participating in voluntourism (see McGehee 2014 and Wearing and McGehee 2013a, 2013b for reviews of the evolution of this field of inquiry). Several studies focus on the potential of voluntourism to increase social, political, and environmental consciousness and participation; to heighten intercultural awareness and understanding; and to foster personal development among participants in these travel experiences.[15] In general, voluntourists are motivated by altruism, "giving back," and "making a difference" in the lives of local peoples (Caissie and Halpenny 2003; Wearing 2001).

While there may be virtuous reasons to support the expansion of voluntourism opportunities, there is also evidence that such experiences can have negative impacts at the local level. For instance, a growing tendency of volunteer tourism programs is prioritizing the needs and desires of tourists over locals. In volunteer tourism situations, volunteers are often relied upon to carry out labor that could be accomplished through the hiring of locals. The use of volunteer labor over local labor means that the work completed can be subpar and unsatisfactory, as volunteers frequently do not possess the skills necessary. Further, the presence of free labor may disrupt local economies and promote a cycle of dependency (Guttentag 2009; Palacios 2010).

Scholars have also observed a persistent "othering" of host communities pervades voluntourism contexts, which often includes a rationalization and romanticization of poverty that follows a discourse that suggests hosts are "poor-but-happy" (Guttentag 2009; Simpson 2004; Sutcliffe 2012; Wearing and Wearing 2006). Such effects are heightened by the fact that volunteer-sending organizations often do not engage in or encourage critical reflection about poverty and global inequality (McGehee 2012; Sutcliffe 2012). Marketing material frequently reinforces a sense of "otherness" onto host communities, portraying western volunteers as bringing educational, material, and emotional aid onto host communities (McGehee 2012; Raymond and

Hall 2008; Sutcliffe 2012). Such commoditization of experiences with "others" has the effect of reinforcing the dominant position of the Global North and limiting one's ability to see positive change within the host community (Guttenberg 2009; Simpson 2004; Sutcliffe 2012).

Yet, voluntourists themselves articulate and project a distinct identity from routine tourists. Voluntourists separate themselves out from the banality of mass tourism and the destructive nature of package tours, because unlike the hedonistic, self-absorbed mass tourist, the voluntourist is driven by social and environmental concerns (Butcher 2003). And voluntourism marketers have embraced this distinction, frequently marketing voluntourism experiences as those in which the more "savvy, resourceful, cultured, sensitive, spontaneous, adventurous and creative" traveler engages (Vrasti 2012, 7). It is possible that volunteer tourism experiences will trigger increases in social activism and improve "global citizenship" as participants become more knowledgeable about global issues and desire to become involved in solutions and change (Butcher and Smith 2010; McGehee 2012; McGehee and Santos 2005). However, research shows that unlike global volunteers of the past, today's volunteers are focused less on cultivating a global citizenship to spearhead their involvement in social activism and more on volunteering as a training ground for future professional employment opportunities created by their accumulation of cultural capital (Butcher and Smith 2010; Heath 1997).

Drawing on research that characterizes members of Generation Y as open-minded, multitasking challenge-seekers with a strong work ethic and respect for cultural and ethnic diversity, Kevin Lyons and colleagues (2012) consider how volunteer tourism is a form of gap year tourism that embraces neoliberal ideology.[16] The cultural capital and entrepreneurial competencies gained through volunteering may be translated into economic capital upon return home (Heath 2007; Lyons et al. 2012; Simpson 2005).[17]

Volunteer tourism can be understood as about the creation of individual identity, and the trip itself becomes part of the individual's "biography" rather than a trip about the host community or society visited. In other words, the experience of voluntourism helps shape and create the volunteer subjectivity. I speak of emerging subjectivities here rather than identity construction, as subjectivity is what our past experiences make us into, whereas identity can be understood as what we are or wish to project. However, both elements are extremely important to conservation volunteers: the projection of a particular identity and the building of the subjectivity of a future global change leader equally contribute to the making of the multispecies conservation volunteer described in this book. Today, ten years after Lyons and colleagues' observations, this "biography" that emerges through voluntourism is also now informed by what makes for a good image—or imaginary—on social media. "It's all about the gram," after all.[18]

On Methodology, Coloniality, and Collaboration

The data I draw on for this work come from several short-term ethnographic fieldtrips (4–10 weeks each, totaling approximately a year) across three major time periods (2001–2, 2011, 2016–20), and twenty years of sustained study of secondary sources and media coverage of the island. The 2001–2 fieldwork was conducted with the intention of comparing tourism development on the island of Utila with that on the north coast in writing my dissertation. That proved unwieldy at the time, and as a result I am only now publishing some of that data.

The 2011 research was dedicated to exploring the relationship between conservation and tourism development on the island. It was at this time that I made an intentional effort to develop relationships with the island's conservation organizations and to learn more about their volunteer programs.[19] I maintained continued communication with the island's conservation organizations after the 2011 summer fieldwork and collaborated with them all at one point or another on web-based surveys.[20] Since 2011, I have also been tracking and analyzing the organizational marketing materials for all of the island's conservation organizations and tourism-related websites and blogs. These materials inform the chapters describing the multiple frames through which Utila is imagined, especially chapter 2.

Starting in 2015, several other actors have been involved in data collection. The study of volunteers was embedded into a course I cotaught with a geographer, with students engaging with conservation professionals to design, carry out, and analyze results of a survey of volunteers (Brondo et al. 2016). Additionally, my good friend and colleague anthropologist Suzanne Kent and I have been working with Honduran conservation professionals in Utila on several projects focused on collaborative conservation and culture change. One collaboration enabled a team of four conservationists (i.e., representing the range of conservation organizations on the island) and four anthropologists (Kent and myself plus two graduate students) to jointly explore the relationship between local cultural values and conservation objectives (Kent and Brondo 2019).[21] A second project enlisted islander stories about their past multispecies relations in the redesign of the local K–12 environmental education curriculum. This project launched as I completed the present book. Several interviews from the 2016–19 joint fieldwork inform this book. All ethnography and interpretation are my own.

Ethnographically, over the years I spent my time "hanging out deeply," both virtually, in the many online mediums that bring Utilians and those who work, study, or play on Utila together, and in person, in islander shops, homes, municipal buildings, volunteer barracks, and boats; in and under the sea, deep in the red, white, and black mangroves, on top of hillsides, and along the shores. I was pregnant with my now twenty-one-year-old son the first time I visited the island, and my husband and I would travel there regularly for the first two and half years of his life while I pursued dissertation research on Garifuna land rights on the north coast. It was during this time period that we also spent slightly more than two months on the island collecting data on tourism development through audiorecorded interviews and ethnographic engagement.

Almost all of my onsite ethnographic work in Utila—unlike my other Honduran fieldwork—was completed with at least one member of my family present in the field. This meant my husband often operated as a research partner, and in 2002 and 2017 he assisted with data collection. My children also regularly participated in volunteer activities, and in 2016 my then thirteen-year-old daughter even kept a field journal.

While scuba certified, my husband, DJ, and I never really took to the diving lifestyle on the island, preferring, as some Utilian women say, "to keep both feet on the ground." Therefore, while nearly all of the island's visitors would be off on boats and underwater during the days and then in the bars at night, by contrast, my husband, son, and I would be on the island's few streets, in locally owned businesses, on docks, porches, or homes, hanging out. We did not fully appreciate how unusual this behavior was for foreign arrivals until 2011, the first time since 2002 that my family returned with me to the field. Nine years had passed for them, and it was my daughter's first trip to Honduras. Within moments of our arrival, a local hotel owner said to DJ and me, without our prompting a memory: "I remember you; you used to ride up and down the street with your boy on the bicycle." This was the only bicycle on the island at the time that had a baby seat on the back, loaned to us by the mayor's wife, a sign of the community care that permeated the air of this small place. Something of this nature happened again in 2020 when I met one of the island's Old Heads for the first time. Upon learning that I was an anthropologist, he immediately recalled meeting DJ in 2002, sharing with great specificity the conversations he had with my husband at the time, who he recalled was "on the island with his anthropologist-wife." He remembered loaning us his copy of Richard Rose's book (featured at the start of this book) and his delight that we were on the island to learn more about the people living there and not just the species under the sea.

Doing research with volunteers was like hanging around with my children's friends: the vast majority of volunteers were just a couple years older than my son. I suspect they were on their best behavior when I was tagging along, even though they served in the role of teacher to me. I am also older than the majority of conservation professionals on the island. Utila is one of the first rungs on the ladder for those seeking to climb in Honduran conservation circles. Employment in Utila's conservation organizations is also fairly unstable, often dependent on soft money generated through grants or volunteer fees. These conditions mean that my partner organizations experience extremely high rates of turnover and that I am continuously working to build rapport with new organizational leaders and staff. Because my work on the ground is limited to short-term visits between semester teaching and administrative obligations, I often return to the field with an entirely different set of organizational actors each season. As well, in tourism contexts, people can forget you and you have to reintroduce yourself—as one acquaintance said after it took him a moment to recognize my face: "I'm sorry. Of course I know you! It's just that I see a lot of white girls come through here."

Due to my background in feminist political ecology and feminist methodologies, I approach fieldwork with careful attention to power, including my own. As I think about power in my work, I am reminded of Juanita Sundberg's warning that we must be ever-aware of the ways in which we reproduce a one-world world through the centrality of Eurocentric knowledge and enacting universalizing claims in our own work (Law 2015; Sundberg 2013). We must acknowledge and confront how fundamental the colonial experience is to the way in which we think and do our work. At the start of this book, in my thumbnail description of Utila, I employed colonial naming conventions, scientific taxonomies, to describe the natural life on the island, and I placed a local term "swamper" in quotations. I could have chosen to edit that out before sending to press, but I chose to keep it within the text because it serves as a reminder of the need to continuously question the reverberations of coloniality in my work. As you read the history and narration of Utila's past, you will notice many of the sources come from white men, settlers who are also authors, or early anthropologists who worked in the region. This pattern holds true from the very opening selection in the book, which begins with Rose's description of the island, through the reliance on early archaeological, ethnographic, and historical geographic studies in chapter 1 (fields that were dominated by men), to media publications written for tourist audiences presented in later chapters. These perspectives are written from a one-world-world ontological frame. I include them throughout while working to poke holes in these narratives, adopting the decolonial methodology offered by feminist political ontology (Anzaldúa 2002). In so doing, my hope is we can imagine beyond the patriarchal culture defined by values, actions and emotions of competition, hierarchy, power, domination, appropriation, and accumulation and move toward what Humberto Maturana and Gerda Verden-Zöller (2008) call "the biology of love." This perspective values inclusion, relationality, collaboration, understanding, and the "act of emotioning" and a "social coexistence based on love" (Escobar 2017, 13). Thus, while my interlocutors were and are largely informed and operating within a one-world-world ontology, my goal in this book is to open possibilities for other ways of imagining Utila's past, present, and future—in other words, to attend to a pluriverse.

Another extension of coloniality is the ability of some human bodies to move freely across the globe—myself, my family, and the thousands of dive tourists and conservation volunteers included—who are engaged in various, emergent relationships with each of Utila's many species. I am keenly aware of the contradictions in my movement and engagement with this global tourism industry and environmental change, as well as the ways in which my own positionality, as a middle-class, middle-aged, white woman, has shaped this work. Over the nearly two decades of visiting and thinking about Utila, I have climbed the academic ladder from student to a tenured full professor. On my last few trips I came with resources to share with the local conservation organizations, supported by two grant programs. I also came with a publication record that included critiques of conservation efforts in Honduras, which could have made Utila's conservationists skeptical of working with me. Instead, they were open and collaborative, and in turn I wish to show them the same respect and care in this work.

My work should not be read as demonizing conservation research organizations and the volunteers that work with them. The researchers I met have great passion for the work they are doing. They see their work as critical to the survival of our collective worlds, or else they most certainly would never probe or kill any other living species. These are people who relocate tarantulas and scorpions in their Utilian residences rather than exterminating them. (Meanwhile, my husband, children, and I developed elaborate workarounds to avoid our own resident tarantulas at all costs, even if it meant sleeping on a bench in the kitchen.) Thus, I am not writing against conservation voluntourists in this book but instead writing to try to understand them a little bit better, for a couple reasons. First, my encounters with conservation research volunteers in Honduras have drastically increased over the past two decades of exploring the relationship between conservation, development, and rights to local resources on Honduras' coastline and island communities. Volunteers in conservation research are truly unavoidable if you want to understand the state of local conservation efforts and how such efforts impact local people. Second, I want to also understand a bit more about myself and other humans like me. What motivates us to engage with other creatures, and in particular with endangered and endemic species in distant places? And yet, despite all our interest in nonhuman beings, other beings are vibrant and entangled in making their own worlds, sometimes in relation with us in ways we may yet to fully comprehend.

Outline of the book

The forthcoming chapters draw on political ecology insights to unveil the entangled lives of humans and other beings confronting and immersed within an expanding global economy based on affect. A political ecology framework necessitates attention to the ways in which affective encounters are shaped by the particular geographies and histories within which they are embedded. Chapter 1 provides the historical context for understanding contemporary conservation voluntourism on Utila. In this chapter, I trace Utila's complex cultural makeup back through the Spanish enslavement of its earliest inhabitants, the Indigenous Pech, to the hundreds of years of armed conflict with Spain as the island remained under British control and served as site for British pirates to hide their plunder (Stonich 1999), to the eventual surrendering of the island to the government of Honduras in 1860. Rather than presenting a unilineal narrative, this chapter presents alternative histories that analyze the relationship between the legacy of colonialism, migration, dive tourism, and today's conservation voluntourism that is arising to address the degradation caused by capitalist expansion. The chapter is punctuated by the telling of the island's history through the words of Utilians, through the few books (nonfiction and fiction based on historical fact) and pamphlets written by Utilians (Jackson 2003; McNab n.d.; Rose 1904; Smith 2013), and oral histories told to me.

Chapter 2 traces the rise of dive tourism on the island of Utila and how it transformed from a sleepy fishing community into a global destination for backpackers seeking drinking, drugs, and dives. Igoe observed that W. E. B. Dubois's (1915, 712, quoted in Igoe 2017) famous quote about white man's privilege "to go to any land where advantage beckons and behave as he pleases," while those of color are increasingly confined to parts of the world where life is more difficult, applies remarkably well to the global tourism industry. Most certainly, it applies to Utila, and is captured in the images trafficked on the Internet about touristic experiences on the island. The title of the chapter— "If You Come to Utila, You Can Do What You Want, and You're Never Gonna Leave"—reflects one of such imaginations, from a YouTube video gone viral that showcases a misogynistic and party-obsessed island lifestyle. This narrative of island life stands in stark contrast to the conservative perspectives and practices of islanders and mainlander migrants.

While changes to the island have been accelerated by global media fantasies (e.g., YouTube celebrity status of the island's "Dr. John," HGTV's *House Hunters International*), other global forces of capitalism are equally responsible, including neoliberal economic policies that drastically transformed the country's agricultural, fishing, and tourism economies. Policy shifts beginning in the 1980s and carrying through the early new millennium significantly altered local livelihoods and migration patterns, exacerbated in more recent years with political instability and intensely high rates of violent crime and corruption on the mainland since the 2009 coup d'état.[22] The unfettered—and largely unregulated—growth of tourism has made for an uneven and a volatile ecological and social environment, with inequalities in access to and management of local ecological resources, as well as has produced a growth in conservation-related research.

Chapter 3 tracks the growth of conservation efforts on the island and the connection between volunteering and conservation. In this chapter I demonstrate how, due to the inequalities in relation to caring (affect), multispecies conservation voluntourism is having concrete and tangible impacts on both local ecology and social relationships to other species. To do so, I consider the ways in which Utila's protected area boundaries are contested both discursively and materially. Materially, I present the ways in which geophysical spaces have become encapsulated and patrolled within Utila's protected areas, alongside the movement of species beyond bounded territories, and the allowance of some categories of people into bounded space, while keeping others out. Figuratively, I explore how contestations over boundaries are closely tied to discourses and actions of "protection." Sometimes these acts of protection are codified into law, resulting from volunteer-led conservation research, which then produce and require different multispecies interactions.

In closing the book, I consider how protected area conservation and development work have been deeply transformed by multispecies encounters and inequalities, in relation to caring, that are produced through the emergence of an affect economy. The final chapters, chapter 4 and the conclusion, are a response to the charge of the new materialisms literature to consider worlding practices that recognize multispecies entanglements and the possibilities for life in colonial-capitalist ruination (Collard, Dempsey, and Sundberg 2015). Emerging in Utila are new ways of being in a landscape drastically transformed from its transition from a small fishing and agricultural community to one reliant almost entirely on dive and conservation research tourism. Affect is produced between locals and short-term volunteers, who have become a steady presence and are warmly embraced due to the small-town nature of this community; it is found between species as they persist living in fractured ecological corridors, as humans exhume and relocate turtle nests to increase chances of survival, as well as when mangroves are cut down and infilled to make room for mainland migrant laborers who see this emerging economy of affect as an opportunity for life in a precarious and unstable country.

Chapter 1

"And the Sea Shall Hide Them"

Utila's Cultural Landscape and Multispecies Entanglements

Take part in a living story. Go back in history, about 350 years and I will show you the evidence that Robinson Crusoe's Island is in fact the picturesque island of Utila... This story has been passed down through my family.

The English pirates had begun attacking the Spanish ships laden with gold and silver that were bound for Vera Cruz, Mexico and Cuba. The Indians living on the islands off the North Coast of Honduras were giving provisions and support to the pirates. They became such a nuisance to the Spanish route that in the year 1640 Spain sent an armada, (a flotilla of armed ships), to clear the islands of it's [sic] inhabitants. The Indians were dealt with firmly. Many were killed and those captured were taken to Mexico and Cuba to work in the mines.

In the year 1659 on September 30, [t]here was a shipwreck during a hurricane and the sole survivor kept a diary of his experiences on the island for 27 years. It has become known throughout the world in the book "Robinson Crusoe" by Daniel DeFoe [sic].

The location of the actual island has puzzled readers for centuries. Robinson Crusoe left a buried treasure before leaving for England on a Jamaica ship. The island location was camouflaged by purposely mentioning Trinidad and the Orinoco River in Venezuela.

Much has been said throughout the past 270 years about Daniel DeFoe who wrote the original "Robinson Crusoe." No one has been able to locate the actual island on which this incredible story has taken place.

"And the Sea Shall Hide Them"

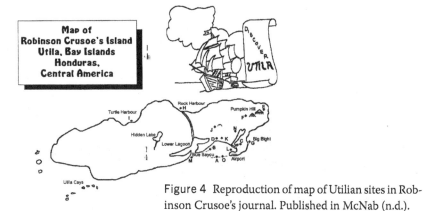

Figure 4 Reproduction of map of Utilian sites in Robinson Crusoe's journal. Published in McNab (n.d.).

On the map ... you will see how the Island of Utila in the Bay Islands of Honduras relates to the diary of Robinson Crusoe. (McNab n.d., 1–2) (see figure 4)

Shelby McNab—local historian, TV anchor, author, and herbalist—has told this story many times. McNab is an Old Head islander, descendant of one of the first white settler families, and author of a pamphlet on Utila's history with a map corresponding to the fifteen sites documented in Crusoe's journal. Each site now has an exhibit in which tourists can read excerpts of the novel *Robinson Crusoe* against the interpretation of Utila at the time (see figure 4). McNab details Crusoe's journey on the island, tracing his shipwreck and trek through the island's fringing reef and into a small land mass in the swamps on the east, which Crusoe immediately found unsuitable for habitation. This uninhabitable swamp is now home to some of the poorest Hondurans. The neighborhood is known as the Camponado (figure 5), and it emerged in the 1990s when the local government approved a program to sell plots in mangrove wetlands to the island's poor and mainland migrants. The name Camponado comes from two Spanish words: *campo* (field) and *nado* (swim), named as such because in the dry season the swamp would dry out so much that children used to be able to play out on the field and would go to the lagoon through the mangroves to swim.[1]

Camponado, the neighborhood, began as a shanty development with substandard infrastructure and services, leading to a series of associated health concerns (e.g., water-borne illnesses and skin conditions). The community

Figure 5 Camponado, 2020. Photo by Josely Turcios.

has continued to grow as more and more mestizo mainlanders move to the island in search of labor opportunities. There is now electricity and paved paths throughout much of the community. While some areas of the Camponado also have improved housing, much of the development continues to be substandard and environmentally hazardous. This mangrove swampland is a prime example of a relational ontology in its constant state of emergence, relationality, and dependence on a multiplicity of life-forms (water, minerals, salinity, etc.), and its transformation into a housing development is a manifestation of a one-world world, an ontology of human dominance over nature (see also Escobar 2017; Ogden 2011).

From today's Camponado, Crusoe travels back to where he first washed ashore and digs a cave in the hillside. He later adds a couple tents and names this his castle. Today, an unfinished five-story hotel resides in this location (figure 6).

Each day Crusoe would travel to the top of Pumpkin Hill, on the other side of the island, the location where I sat at the start of this book, where he

Figure 6 Photo of Robinson Crusoe exhibit. Photo by author.

could scan the sea for shipping vessels. McNab writes, "From here Robinson Crusoe sees the mountains of the mainland. He also found that there were lots of turtles on this side of the island" (n.d., 3). Historical accounts from the late seventeenth century through the 1800s often referenced the abundance of turtles, with geographers reporting back to England that the area surrounding the Bay Islands "swarms with fish and turtle, both green and hawksbill" (Roberts 1827, 275–77; qtd. in Davidson 1974, 74), and the limited passages through the island's surrounding reef were "being mostly made use by the turtlers" (Jefferys 1762, 51, 100; qtd. in Davidson 1974, 59). Turtles and turtling have long been a figure in Utila's economy of affect, predating and then intermingling with the island's agricultural production. Up until just a few decades ago, island children would keep baby turtles in a crawl (a fenced-in spot in the ocean close to one's home where fishers transferred their catch from their dories) off their docks. Marion, descendant of one of the Old Heads, recounts:

> I remember growing up, we would have turtles for pets. Turtles you'd pick up out of the seaweed. Your parents would then get an old icebox and drill a hole through that, put a couple rocks inside of there and plug that hole with a piece of cork, and walk out to the dock every day or every two days and pick up buckets of saltwater to put freshwater in there for the turtles, and once they turned three, six months you had them, then you'd paddle out in a dory and set them free.

With the rise of conservation in the 1990s, a new message emerged to accompany turtle care. No longer were children simply scooping up turtles from the seaweed and caring for them as pets; in that moment, a former Peace Corps worker introduced a program that would help transform local views of turtle from "food" to "friend." McNab, who descends from a family of turtlers, joined the man in rolling out this program through sponsorship by the BICA, transforming his inherited familial identity from a harvester of turtles to guardian of them.

According to local families who participated in the first turtle project,[2] it involved caring for around ninety hatchling turtles and targeted children who were responsible for feeding and caring for them until they were old enough to be released. They were then warned against eating turtle meat because it could be from a hatchling that they had "raised." Here we see a transformation in care—earlier, perhaps, families cared for hatchlings to raise them until they were ready to be released into a community, where they could reproduce themselves and contribute to the numbers that could be eventually harvested for consumption. Now all turtles are off-limits to the local population unless you are caring for them so they might survive against the odds of human tastes for turtle eggs, turtle meat, or disposable plastics.[3]

Today, human turtle consumption has transformed from consuming for sustenance to consuming an experience or encounter by a tourist or conservation volunteer. In time, such care became organized under a formal program associated with BICA. The organization continues its management of the turtle population, now under the science of conservation and through careful patrolling and monitoring of sea turtle nesting sites.

Below Pumpkin Hill, where Robinson Crusoe hiked to scan the seas, lies a small beach where the hawksbill and loggerheads come to nest. Since 1992, Pumpkin Hill Beach has also become a site for conservation science, and its nesting sites are now regularly patrolled and protected by BICA staff and volunteers. In 2018, foreigners from the United States purchased land along the shoreline, and by the summer of 2019 a handful of new homes dotted the coast. Some of the land was sold by Mr. McNab himself. At the time of writing this book, the impacts of this construction on turtle nesting activity were uncertain, but the homeowners were collaborating with local conservationists to monitor turtle activity and to reduce light and noise pollution in the area.[4] One new landowner from the United States was helping by bringing garden fencing from Home Depot to mark nesting sites, replacing the earlier stone markers that would often be trampled upon by people walking the shoreline unfamiliar with their significance (figure 7).[5]

Figure 7 Garden fencing surrounding turtle nesting site on Pumpkin Hill Beach, December 2019. Photo by author.

Nearby Pumpkin Hill sits Brandon Hill Cave, also known to the Kanahau researchers as "the bat cave," for this limestone cave system is home to an array of bat species present on the island. Kanahau researchers are engaged in cataloguing the range of bat species and estimating the population size of bat colonies within the cave system. Among the bats in the cave are "duppy" spirits or ghosts, malevolent forces that come out at night to haunt people. Duppy (also spelled "duppey," "duppie," or "duppa" by islanders) spirits, a word with West African roots, are common throughout West Indies Caribbean folklore and were brought to Utila with their Cayman settlers, as well as from Robinson Crusoe, who passed through the West Indies before becoming shipwrecked on Utila. McNab reminds us of Crusoe's encounter with a duppie in Brandon Hill Cave: Crusoe encounters a ghost, and he writes: "I saw two broad shining eyes of some creature, whether devil or man I knew not, which twinkled like two stars, the dim light from the cave's mouth shining directly in, and making the reflection" (Defoe [1719] 2007, 227) and "I saw laying on the ground a most monstrous, frightful, old he-goat, just making his will, as we say, and gasping for life, and dying, indeed, of mere old age" (228).

Duppy stories are common on the island, and McNab's pamphlet suggests that Crusoe was one of the island's earliest duppy storytellers. In 2018, Winnie, a phenotypically brown "white" Utilian woman who grew up in the 1990s describes the duppas that haunted her as a child: "Our duppa stories! They were the best. We used to sit at night when the lights went off, everybody sittin' on the steps and you would hear these duppa stories. And you was a *woman* to pass that big gutter, a woman or man to pass that big gutter once you heard about the masked man livin' under Miss Rosa's house floor." Adults used to tell stories about the masked men living under porches, and in the dark, children could see big round faces and dark eyes staring at them. Nowadays Winnie believes what she was actually seeing were boat buoys hanging under people's homes. But back then, the masked men haunted you. The duppa stories had Winnie so frightened that she used to be afraid of her own shadow and would run as fast as she could to try to escape it. Today she reflects, "These were just stories to keep us home a little bit more."

In Winnie's case, duppa stories were used as a way to keep children safe and close to home. But they also appear in contexts where islanders had limited knowledge—and no familial connection—to past human inhabitants. In *Black Chest: A Vault of Historical and Cultural Knowledge Stored by the*

Black English Speaking People that Reside in the Bay Islands and the Honduran North Coast, a 2013 book detailing the oral histories of black English islanders, A. E. B. Smith (2013, 110) shares, "African beliefs are still alive in Black English speaking Bay Islanders oral culture" with islanders using the term "duppey" to refer to all different types of spirits of the dead. Often duppies appear alongside tales of pirates and Indigenous people intermingling. Smith (2013, 110) writes:

> Many stories are told of Pirates' Gold and Duppies (ghosts). It is frequently rumored that after pirates raided Spanish ports and ships, they returned to the Bay Islands to bury their treasures and goods; just inland from the bights and bays of these isles. However, rumors also state that while the pirates were securing their booty the aborigines would watch where they hid it from the crest of hills throughout these islands. When the pirates left, the Payas would dig up the treasures and hid [sic] it somewhere else, usually at the top of hills such as Dixon's Hill [on Roatán], Spyglass Hill [on Roatán] and Pumpkin Hill.

According to McNab, duppy stories are why islanders are afraid to enter Crusoe's cave nearby Pumpkin Hill. McNab himself has entered on several occasions; in fact, he owns the property where the cave sits. Inside McNab found jade and a gold cup, likely part of Crusoe's treasure, he assumes, taken from the Spanish ship that had washed ashore in an easterly storm on what is now referred to as Big Bight (McNab n.d., 7). Rose (1904, 147) also mentions this cave at the top of Brandon Hill in his 1904 account of the island's history: "Many years ago, a golden goblet, a rusty sword, and a golden crucifix, were found in the cave. They were brought out and sold by the finder to Mr. Whitefield, one of the merchants in the island. The goblet was finally sent to an exhibition that was held at Comayagua, the former capital of Honduras." For Rose, the origin of these items is not as clear as they are to McNab. Rather, Rose (1904, 147) states, "There have been many speculations as to how these articles were placed in the cave; but it is not likely that the truth will ever be known." It is unclear if McNab is referencing finding the same goblet in his lifetime as the one that Rose states was sent to Comayagua in the late 1800s (McNab was born in the 1940s).

The above accounts from islanders reveal the entangled history of Utila, with narratives of exploration, enslavement, and conservation science. Contemporary descendants of settler families work to make sense of their past through narratives of spirits intermingling with material remains and speculations about ancestors based on artifacts islanders have encountered in their landscape.

Evidence from the Past

In 1931, Junius Bird led an archaeological expedition of the Bay Islands for the American Museum of Natural History. The Smithsonian Institution carried out a similar survey in 1933, and the combined results of these two expeditions are reported in a 1935 Smithsonian publication authored by anthropologist William Duncan Strong. In this section, I present excerpts from the Smithsonian publication to illustrate the ways in which Utila's landscape was transformed through the one-world worldview of colonists and settlers dedicated to capitalist accumulation and managed landscapes. Brandon Hill Cave, or Robinson Crusoe's Cave in McNab's account, was one of the sites they surveyed.

> On a rocky ridge not far to the southwest of Pumpkin Hill is a deep cave of which one hears as soon as relics are mentioned to anyone from the islands. According to stories from many sources this remarkable cave penetrates down to salt water, or according to another version extends all the way to the southeast shore of Utila. A golden cup, a rusty sword, and a crucifix, as well as a cache of rifles, are among the finds reported from here, and the gold cup incident at least is apparently true (see Rose 1904, 147). Incredible tales of magic and buried treasure float from mouth to mouth around the islands, and a folklorist would have a rich field for investigation among both whites and negroes. While no "treasure" finds other than this one gold cup (apparently looted from a church) ring true, the rumors have their inevitable concomitant in a senseless destruction of Indian sites in the fruitless search for gold.
> Brandon Hill Cave is a beautiful place, and wherever its winding limestone passages may actually lead, they are extensive enough to satisfy the most ardent cave explorer. . . .
> Bird explored this cave until the descending crevices became too small to permit passage. The material which he obtained indicates that at least the main chamber was occupied by the aborigines. This collection includes

some 21 sherds from the shallow dirt floor just inside the entrance, all of which had been turned over by treasure hunters. (Strong 1935, 30–31)

The passage continues, describing each of the sherds in color, size, and thickness. Other finds are detailed, including a fragment of a stone bowl, a sharp-edged celt (i.e., a prehistoric weapon) made of a conch shell, an obsidian knife, several large land snails with perforations, various shells, glass bottle pieces, animal bone, and a number of small human skull and other bone fragments. Strong (1935, 32) notes, "The later are of considerable interest but unfortunately no likely places for more complete burials could be found in the cave."

Utila is full of evidence of the past, but contemporary archaeology excavations are virtually nonexistent. Recovered artifacts and managed landscapes suggest a vibrant culture and community, but contemporary residents know very little of the island's earliest residents as the population was decimated before their ancestors arrived, and so they draw their own conclusions based on their encounters. Rose (1904, 38–39) reflects on observations from the 1870s:

Small earth-mounds and rock-mounds have been found in many places in the islands . . . my attention was drawn first to some of those that are to be met with in the cocoanut-gardens on the south side of the islands . . . I thought at first that the owners of the gardens had attempted to clear them of the stones and so formed these ridges. But upon inquiry I found that neither of the oldest settlers knew anything about when or by whom these mounds had been formed. I was, therefore, convinced in my own mind that the work had been done by the Indians. Evidentially the builders of the mounds knew something about order and were painstaking . . . One of the mounds at the north side of the island was opened June 12, 1897. The diggers after hidden treasure supposed to have been buried by the pirates. Instead of precious metals, however, the seekers found in this mound many curious articles.

Rose had the opportunity to observe many items that emerged from this excavation, as another early settler, Dillard Whitefield, participated in the excavation and shared the artifacts with him. Rose observed, measured, and described the water pots, urns, arrowheads, coral beads, ivory hatchets, and other objects that were removed from the mound. Today, several of these

"Indian relics," as Rose called them, reside with an expat on the island who has been collecting artifacts on the island with the intention of eventually opening a cultural history museum. The museum plans have been in the works for at least a decade, that I have known of, and now the museum appears to remain eternally delayed.

As well, a local artisan sells pottery sherds with paintings of Utila's landmark sites; the source of these sherds is also likely from the old "Indian cemetery" located on the north side of the island or from midden pits and other looted sites in an area called "Bambu," located behind the baseball field and toward the neighborhood of Jericho. Much of the land around Jericho has since been transformed into farmland and cattle pasture. Rose (1904, 41) describes some of Indigenous sites that he and other early settlers encountered in their time:

> Reader, let us go over some of the ancient Indian roads in the island. Near the centre of the island Stuart's Hill rises above the lowland. From the top of this little hill the sea is visible almost all around. The lowlands and the lagoons spread out in lovely vista beneath the spectator. To the eastward a little northerly of this hill, at a place called Bamboo, is one of the Indian's ancient burial grounds. From that old cemetery a road paved with stones leads to the top of Stuart's Hill.

In fact, the archaeological record shows evidence of residential, ceremonial, and burial sites on Utila, including around forty acres at a place called "80 acre," a residential area in elevated locations on gradually sloping land that ended at the beach, as well as the burial site on Stuart's Hill (Davidson 1974, 20). Strong (1935) reported that 80 acre was probably one of the main habitation sites connected by the stone paths that Rose mentions. By the time of the 1933 expedition, islanders had already dug up much of the burials, leaving just "some scattered human bones and teeth along with sherds and other refuse" (Strong 1935, 35).

Pillaging was commonplace and continues to be discussed as a favorite pastime.[6] In 2019, Clara, a middle-aged Cayan woman, tells me how her uncle used to take her "nature-walking"—"that's what he called it"—to find skeletal and other material remains among the lava rock along iron shore and on the Cays. She said her father, a carpenter, found entire bones, "hands, elbows, feet," when constructing homes, and they didn't know what to do with them so they simply threw them away. Multiple islanders reported finding human remains at the old "Indian cemetery" on the north side of the island, as well as in the Cays.

The Smithsonian excavation report cites islander accounts of burial sites in the Cays and other areas, but their expedition only included the north side burial site called "Black Rock Basin." A considerable amount of remains, including human remains, were found here, even after looters had come through. Black Rock Basin and Jack Neil's Cay are both known among islanders to be haunted. Stories abound of spirits haunting those who walk upon the grounds of Jack Neil's Cay. Clara shares, "That cay was the Indians' graveyard. But they weren't proper burials; just remains of everyone. Those days people were very wicked people. They'd fight with machete, rocks, all kinds da of things." Maybe it was a war zone, she ponders. It may well have been, but perhaps the wicked people were not the ones whose remains were left littered across the landscape.

Reading the Smithsonian report through the lens of political ontology makes visible the accumulation by dispossession associated with the decimation of the Indigenous population and establishment of coconut plantations, creating new landscape assemblages that current generations of Utilians work to apprehend. In describing Black Rock Basin burial site in 1935, Strong (1935, 22) writes:

> [To reach the site] we rowed in a small boat to the west end of the basin. Landing here at a small native plantation, we walked west about half a mile along the low shore line, which consists of rough coral rock interspersed with occasional bits of sandy beach. The entire shore line is heavily grown up with large coconut palms. Within half a mile of the west end of the basin potsherds, conch shells, and other traces of aboriginal occupation appear, and these extend to the west with varying intensity for at least half a mile and perhaps farther.

While these archaeological accounts go into great detail regarding the artifacts located at the site, the point I want to underscore from this excerpt and the earlier excerpts is that the transformation of Utila's landscape was brought on through the one-world-world ontology of colonists and settlers: an ontology of order and capitalist production. We see references to a one-world-world ontology in Rose's (1904) account of the conversion of forests in the Bambu area to farmland and cattle pasture. We even see the one-world

world in how Rose imagined landscape assemblages from the time of the Pech. Recall his statement that "evidentially the builders of the mounds knew something about order." For Rose—and the colonialists, explorers, and settlers who came before and after him—reciprocal relations between people and their natural surroundings are hard to fathom, due to the hegemony of the way of worlding associated with modernity.

Colonialism, Buccaneering, and Indigenous Removal

While there continues to be debate in the archaeological community regarding the precise cultural affiliation of pre-Hispanic Bay Islanders, it is likely that Utila was first inhabited by the Pech, a contemporary Indigenous group believed to have descended from pre-Hispanic migrants from South America more than three thousand years ago, who relied on hunting, fishing, and small-scale cultivation (Davidson 1974; Figueroa, Goodwin, and Wells 2012; Wells 2008). With Spanish contact in the 1500s the Pech of Utila were killed or displaced, and by the time McNab dates Robinson Crusoe's arrival on the island, the Indigenous population had been all but wiped out. Just a few "Caribs" remained, and Crusoe encounters them along the shallow bay close to Utila's old airport (which was an unpaved runway along the shore and is now the site of a private beach and restaurant called "Bando Beach," named after a traditional coconut-milk-based stew prepared with a mix of plantains, bananas, yuca, and other locally sourced root vegetables). McNab writes:

> Shallow bay close to airport is where captives were scarified and eaten. The Carib Indians inhabited the East coast of Honduras and frequently brought captives to Utila in their canoes. After living alone for a longtime, Crusoe rescues a man from Cannibals. He calls the man Friday. Because he rescued him on that day. Friday becomes Crusoe's trusted friend and servant. The term, "Man Friday" has come to mean any trusted servant.[7] Cannibals would bring their captives here after a fight. Evidence of this can be found in burial jars that contain bones, skulls and charcoal especially on the north side of the island. [Here again McNab is referencing the "Indian cemetery".] (n.d., 9)

There is no historical evidence that the Indigenous of Utila were cannibals, but saying that they were served the purposes of the Cuban-based slavers, who declared the Indigenous population as inhospitable, hostile, and opposed to Christianity and enabled the colonialist worlding to unfold in the region. In 1503, Queen Isabella authorized the widespread capture of any Indigenous person designated as "Carib" (cannibalistic), and thus, all that the slave raiders needed to do to justify their raids was to declare an island "Carib" (Davidson 1974, 31; Sauer 1966). It was especially important to reclassify islanders as hostile after the earliest accounts found them peaceful and modest, two highly valued traits of the Christian colonists.

The first accounts of Indigenous peoples on the islands were documented by three historians who accompanied Christopher Columbus on his fourth and last voyage to the New World. These sources became part of Ferdinand Columbus's extensive collection, along with his own writing documenting that journey (Strong 1935, 9). Ferdinand was Christopher Columbus's second son, and he joined this final voyage to the New World at the age of fourteen. It was on this trip that Columbus sighted the island of Bonacca (Guanaja) on July 30, 1502. It is worth quoting the account of the historian Bartolomé de Las Casas at length, as the detail provided reveals the colonial attitude toward the original inhabitants at the time, which then shifts drastically in the next approximately fifteen years. Reflecting upon this first encounter, Las Casas writes:

> Finally, with great difficulties, dangers, and indescribable labor, they arrived and discovered a small island that the Indians called Guanaja, and it had for neighbors three or four other islands, smaller than this one, that the Spanish afterwards called Guanajas, all were well populated. At this island the Admiral commanded his brother Don Bartolomé Colón, Governor of this island, that he go ashore as captain of a boat and get news; he went, taking two boats full of people, found the natives very peaceful and of the type of those of the other islands [i.e., the Antilles], except that they did not have broad foreheads, and, because there were many pines there, the Admiral named it Isle of Pines. (Las Casas, qtd. in Strong 1935, 11)

While the voyage to the Bay Islands was a difficult one, once there, the colonists were pleased to encounter a peaceful population, whom they associated with the native population of the Antilles. The next section of Las Casas' account details the arrival of a "canoe full of Indians" who were returning from a trade journey, likely in the Yucatán. The long excerpt describes the industrious and clever nature of the Indigenous population, with specific mention of their canoe craftmanship and impressive trade networks and the inclusion of a long list of goods from the Yucatán and beyond. These items included a supply of cotton clothing and blankets, swords, knives, hatchets, bells and other medal products, and various foodstuff, including cacao nuts, corn bread, and sweet potatoes. Next, Las Casas takes care to describe how those they encounter "do not dare to defend themselves." In other words, they welcomed the colonists, possibly seeing them as a potential future trading partners. The Las Casas account continues, stressing the modest nature of the "Indians," who covered their bodies in ways the colonists found appropriate.

> Up to 25 men came in the canoe and they did not dare defend themselves nor flee seeing the ships of the Christians, and so they took them in their canoe to the Admiral's ship; and those from the canoe climbing onto the ship, if it happened that their underclothing was caught, then they put their hands in front of them, and the women covered their faces and bodies with shawls as the Moors of Granada used to do with their scarfs. (Las Casas, qtd. in Strong 1935, 11)

Two other historians and Bartholomew Columbus, Christopher Columbus's brother, also wrote accounts of the voyage, each stressing the peaceful and modest nature of the islands' first inhabitants. From this encounter, Columbus took with him one man named Giumbe, to serve as a translator and guide to nearby islands and coastal communities. When they reached an area where he was no longer understood, Bartholomew Columbus writes that he was given presents and sent back to his own country "very contented" (Strong 1935, 10).

Peaceful encounters with the Spanish were short lived, as attention was turned to the islands soon after the slave raids of the Greater and Lesser Antilles had depleted them of their Indigenous population (Strong 1933). From 1516 to 1525 Bay Islanders were subject to Spanish slaving expeditions, and the Indigenous population was sent to Cuba for work in the mines and on sugar plantations. The first licensed slavers were sent on an expedition to the islands in 1516, where they captured three hundred "Indians" and killed any others who resisted. However, once in the Cuban harbor, the Bay Islanders took over the ship and set sail back for their homeland. The Spanish colonists retaliated and sent two more ships to track those who escaped, and they returned with four hundred from Utila and another island (Davidson 1974, 31–32). Additional raids of the Bay Islands continued into the mid-1520s. Population records are scarce, but Davidson's historical review found three sources that show a dramatic decline in population for all of the Bay Islands during this time period. In 1544, 150 houses were documented in Utila. In 1582, a census reported forty married Indigenous people on Utila. Fifty-seven years later, a 1639 census found only twenty-two *tributarios* (tribute-paying) Indigenous people on Utila. While the unit of analysis shifted between the accounts that Davidson could locate, what is clear is a consistent reduction of the Indigenous population over time, much like had been found across Middle America (Davidson 1974, 34–35).

The depletion of the Indigenous population to fewer than two dozen *tributarios* in the seventeenth century was linked to the impact of buccaneering on Spain's ability to control the region. In the seventeenth century, the Spanish had risen as a top trader in American products, drawing the attention of other European nations and individuals, who quickly discovered that they could increase their own treasuries by raiding and plundering Spanish ports and shipping routes in the Caribbean. The Bay Islands were a strategic center, offering easy access to Spanish ports within the Gulf of Honduras, and the islands themselves were brimming with necessary provisions (Davidson 1974, 39–40). French, Dutch, and English pirates established themselves throughout the Bay Islands and raided Spanish ships; they were provided food and shelter at the Indian plantations (Strong 1935, 15). By 1639, Francisco De Avila, then governor of Honduras, was ordered to define a policy toward the Indigenous population on the Bay Islands. And between 1639 and 1641, the Spanish began to make plans for the complete removal of the Indigenous population, reasoning that if they removed them altogether from the islands, then the English buccaneers would no longer be able to find food, shelter, and other provisions on the islands. Hence, De Avila was ordered to "Depopulate the islands—Guanaja, Utila, and Roatán for the great inconveniences they have caused . . . Burn their settlements, as well as their fields, milpas, and cemeteries, and bring the Indians themselves to the land within twelve leagues of Truxillo [Trujillo] . . . the governor is responsible for the passage of said Indians to these coasts so the enemy is prevented from being aided by them" (MS: AGG 1641, qtd. in Davidson 1974, 44). At this time there were approximately four hundred Indigenous people living across the Bay Islands and while the people of Guanaja were considered to be aiding and conspiring with the buccaneers, those on the other islands were seen by De Avila as loyal to Spain (Strong 1935, 15). This is the time period when Utila was reported to be populated by only twenty-two *tributarios*.

While the Spanish were making plans for removal of the Indigenous population, the French, English, and Dutch buccaneers were continuing to disrupt Spanish control of the Caribbean. In the first half of the seventeenth century, Port Royal, Roatán was one of the most easily defended harbors in the Bay Islands and thus became a site of continuous violent struggle. Dutch buccaneers began ravaging Utila and Bonacca (Guanaja) in 1639, and by 1642 English logwood cutters and illicit traders had fortified the harbor at Port Royal. The Spanish had to make several attempts to recapture Port Royal from the English buccaneers, and in 1650, once the English were defeated and the Spanish reclaimed the islands, all Indigenous people were removed. If the idea of depopulation was to counter Indigenous support of the English intervening in Spanish commerce, the plan did not work. Davidson (1974, 44–46) writes:

> With the removal of the islanders the landscape reverted to a more natural state. The burned remains of settlements began to rot, the forests soon grew over former Indian fields, and wildlife, now unhunted, increased. The Spanish plan to depopulate the islands was apparently ill-conceived; instead of ending English interference with commerce in the Bay of Honduras, the depopulation caused the islands to become the object of more serious English plans for settlement.

With English settlement came the advancement of a one-world-world ontology that privileged a plantation landscape of order, monotony, and exploitation. As Escobar (2017, 70) observes, "The plantation form effaces the relations maintained with and by the forest-world; emerging from a dualist ontology of human dominance over nature, the plantation is one of the most effective means to bring about the ontological occupation of local relational worlds."

❖ ❖ ❖

For centuries, colonial narratives have inscribed settler worlds over Indigenous landscapes, attempting to both physically and ideologically erase Indigenous peoples from their heritage landscapes.[8] Postcolonial analyses of historical documents reveal how colonial discourse encouraged the link between race, place, exploitation, and control. Similarly, blackness—like brownness and Indigeneity—was advanced through colonial ideas about race, belonging, and the natural world. For instance, in *Becoming Creole* Melissa Johnson (2018) reviews colonial documents from the mahogany industry in Belize showing how industry documents used language that racialized innate ability (rather than intelligence or creative thought) of "the negro slaves" to find the mahogany trees, emphasizing the animality of the huntsmen and associating African descendants with the jungle. Colonial accounts identified African descendants as "happiest when settled in a swamp and surrounded by mosquitos," who would work just a half hour to furnish food for a couple of days and with no desire for an education (qtd. in Johnson 2019, 34–35). For the British authors Johnson cites, there existed "very few industrious men" in the region (qtd. in Johnson 2018, 35). Colonial writing thus encoded ideas about blackness and brownness with the innate desire to work in the bush, implicit assumptions about who belongs in which places and spaces, and ideas about what are worthy ways of using and interacting with landscapes.

Assumptions about the precolonial socionatural world are reflected in the writing of Utila's settler families, who took with them their Puritan values and imprinted them onto the landscape that they encountered in the late 1800s. Writing at the turn of the twentieth century, Rose (1904, 43–44) imagined:

> that Utilla [sic] was once densely populated, evidence on all sides betoken. Henceforth we shall try to think that the people, who, long, long ago, inhabited this island were a race of intelligent beings of a high morality and civilized . . .
>
> So we shall imagine again the times when the Indians were roaming through the woods in the island, cultivating corn, plantains, and cotton, catching fish in its waters, and making vessels of pottery of various shapes and sizes that served them for all purposes, thus sowing the spirit of energy and industry around them in the island. For we will not think of them as lazybones.
>
> We will think of them too as happy and peaceful until the slavers, armed with licenses from Spain, came and captured and carried many of these perhaps harmless aborigines into slavery in Cuba.

Rose's imaginary of Utila's first inhabitants is one that reveals part of what Sylvia Wynter (2003) calls "Man," a Eurocentric conception of the category of human brought into being during the Enlightenment. This western bourgeois conception recasts prior understandings of what it means to be human to a singular concept of human as rational, law-abiding *Homo economicus*. This notion of Man assumes and naturalizes a racial order, whereby those at the top of the sociopolitical hierarchy are calculating economic actors (Da Silva 2015).[9] Rose interprets Utila's past through this lens, and reads the first inhabitants as having an industrious nature ("they are not lazybones") who transformed the landscape to make productive use of it. Utila's early settlers did not confront Indigenous peoples head-on, so they created an idea of what once was, celebrating the "noble savage" that they imagined through their western, one-world-world lenses.

Anthropological scholarship on landscape assemblages calls into question Eurocentric logics of human domination over and capitalistic understandings of nature, offering instead an approach to conceptualizing landscapes and human beings as relational (Johnson 2018; Kohn 2013; Ogden, Hall, and Tanita 2013; Tsing 2015). Through this lens, socionatural assemblages emerge through entanglements as people and more-than-human entities continuously inform and reshape space, place, and relationality. While the written accounts of Utila's early history are restricted to colonist and settler accounts, this does not preclude us from imagining an in-between-the lines relationality between humans and more-than-humans, and to deduce the ways in which race relations assemble through species encounters and landscape usage. The next section underscores this point in the telling of race relations through historical fiction on the sea and in the mangroves of Utila in the twentieth century.

"And the Sea Shall Hide Them": English Settlement, Old Heads, and Race Relations

Relationships between white and black islanders, and now Honduran mestizos (whom both groups call "Spaniards," a negative epithet that will be unpacked later), cannot be understood without reference to the push-and-pull forces for each to migrate and settle on Utila. As detailed earlier, Utila's Indigenous population was wiped out by the Spanish in the 1600s, and the subsequent hostility between the Spanish and British coupled with significant buccaneering activity kept the island largely absent of permanent settlers through the 1700s and into the 1800s. Postabolition brought new arrivals to the Bay Islands. In the 1830s, whites on the Cayman Islands were outnumbered by the slave population five to one, and fearing that abolition was going to result in the loss of political and economic power, many white Cayman Islanders left for Belize and the Bay Islands. They were already familiar with the area since turtlers from the Cayman Islands often harvested in this region (Davidson 1974, 74). It was at this time that Utila's first white Cayman families settled on the Cays and south side of the island. Significantly, they were fleeing the Caymans for fear that they would lose the economic and political power that came from their social position as whites.

The Cooper family and an American from Massachusetts named Samuel Warren are said to be the first families to settle the Cays. There are eleven islets in the Cays, but only the upper and lower Cays—Pigeon and Howell or Jewel (Cayans use both names)—are settled. In *Utilla: Past and Present*, Rose (1904, 63) refers to the upper Cay as Suc Suc Cay, and other sources also use the name Suc Suc, though I have not heard it used much by Cayan families. At the time of Rose's writing, only the upper Cay was settled; the lower Cay had a chapel, school, and store (1904, 63). The Coopers were one of many working-class British subjects from the Caymans who came to Utila in search of fertile land. Soon after the arrival of the Coopers, other Cayman families followed. Contemporary Old Heads all descend from these first settler families, all of whom identify as white Utilians: the Coopers, Thompsons, Howells, Morgans, Boddens, Diamonds, and Gaboruels. These initial settler families all individually cultivated plantations on the nonresidential Cays (Lord 1975). The main island of Utila remained largely unsettled for a good portion of the 1800s, with buildings limited to dormitories in the east end that housed the Cayans when they were working on their plantations. Temporary housing meant the Cayans did not have to travel back the six miles to Pigeon Cay. The mangrove swamps on the west of the island prevented settlers from planting closer to their residential area.

Slavery officially ended in the Cayman Islands in 1834, and after the four-year "apprenticeship" that enabled former slave-owners to continue to hold former slaves ended, freed slaves immigrated in a second wave to the Bay Islands, settling first in Roatán. Afro-Antilleans began to appear in Utila in the mid-1950, with small numbers documented on the censuses of 1855 and 1858 (William Davidson, personal communication, October 22, 2020). Smith (2013) describes the first black settlers:

Like their white predecessors, they too were primarily from the Cayman Islands—Grand Cayman, Little Cayman, or Cayman Brac—near Jamaica and the ones with Latin sir [*sic*] names from Cuba and mainland Honduras, were all interested in farming. Some of these immigrating black families bore the same surnames as white Utilian families, testimony either to a slave heritage whereby slave names had been adopted in Cayman or to having (at least in one known case) been raised in white families and taking the names of foster parents. (2013, 99–100)

These two waves of immigration—white and black—were inscribed geographically, with residential segregation continuing in Utilian neighborhoods until recent decades. In 2012, during the research for the book *Black Chest*, Smith (2013, 99–100) observed, "There exists a particular social stratification on this island . . . a locally recognized strata based on ethnicity or in other words race." Alongside a photo of Utilians dancing the plat pole in the early 1900s, reprinted in *Black Chest*, Smith (2013, 100) notes: "[This photo] . . . shows the white people dancing the plat pole but all of the musicians were black men. Every important sport event, political, economic position of that era and still today was controlled by the whites. Of all the Bay Islands Utila is considered the most racial[ly] sensitive island even until today."

This "sensitivity" is sensationalized in the historical fiction, *And the Sea Shall Hide Them*, authored by William Jackson, descendent of one of the Old Heads / settler families of Utila. The book dramatizes the real-life events of a 1905 tragedy in which a black islander murdered more than a dozen islanders on board the *Olimpia*, a passenger and cargo ship headed for Coxen Hole, Roatán.

The opening to the novel reveals the deep attachment many descendants of the Old Heads (of which the Jacksons are one) have to the sea and to their British ancestry. The author is William Jackson, a white man born on Utila though raised by extended family in Kansas, United States. Jackson left for school in Kansas at eleven but returned home regularly to see his Utilian family and friends. He opens his novel in the following way:

The verandah with its wing, several rockers and chairs, faced the west, and had an excellent view of the sweeping curve of the island's harbor. Looking down the shore one could see few of the twelve small cays that lie scattered like a following litter just off its Western tip. Directly south, just twenty miles away, stretched the jagged coastline of *Honduras*. On clear days the impressive mountain peaks of that storied land were easily visible. Out there was once part of the Spanish Main, with its history of many violent sea clashes between the ships of Spain and those of Britain. It was peaceful now, here on Utila, and Mr. John Ebank's stable, industrious ways contributed to the island's prosperity. His ancestors came from the British Isles to Grand Cayman and on to the Bay Islands during those turbulent years.

Mr. John and the other inhabitants had a special relationship with the sea just as their forefathers from British seaports had. They knew a great deal about these waters, making their way in them day and night to earn a living. There were days they loved and welcomed its nearness. Other times they feared it. This day, the sight of it peaceful and calm stretching out towards the horizon was a view to be savored.

I had come to visit with Mr. John—perhaps to hear a story—a particular one if I were fortunate. No one had ever told me the full tale of that disastrous voyage in the early part of the century... Some parts of it had been told to me but I thirsted for more. My parents still lived here and were overjoyed to have me home safe from service in the U.S. Army...

I was joyful at being home once more, renewing my ties with parents, relatives and boyhood companions... After arriving I once more bonded with my parents, relatives and former friends. Next, I must visit with Mr. John Ebanks. There was a place in my heart for Mr. John. His home, right on the harbor front, was a haven for my brother and I. We often fished for snapper and silver fish from the doorways of this storage house built over the sea. (Jackson 2003, 12–13)

Even Utilians who spend much of their lives off the island demonstrate a deep knowledge of the marine landscape and its inhabitants, cultivated through a lifetime of seafaring and fishing lifestyles and livelihoods. In this opening, Jackson sets the stage for the story he had long desired to know in full, a true account of a series of murders at the turn of the century, just a handful of years after Richard Rose penned his book about the island.

The fact that Jackson was raised in Kansas, and thus racially formed in the United States, is not inconsequential to his telling of *Olimpia* tragedy. Jackson casts Robert Mcfield as a killer who had long harbored feelings of embitterment and hostility toward white islanders. Years of mistreatment of blacks by whites boiled up and sent someone who all Utilians thought was one of the nicest men on the island on a murder spree. The underlining message of the novel appears to be: you never really know what is going on inside a black man's head, even if he seems to be happily playing his accordion as part of the band for the white dancers. (Mcfield is presented in the book as a well-liked musician who played all the local dances.) Jackson sets this dynamic up by telling us what was going on in Robert Mcfield's head as he stowed away below the ship deck with pigs and other nonhuman cargo:

> I gots to do what I gots to do, he told the pigs. Lots a Morgans on board. His anger was always gnawing at him. He had never gotten over the bitterness when they had refused him a loan. Wouldn't give him a tuppence. Maybe he wouldn't be here this night if they had treated him decent. Dey has hifalutin' ways. Dancin' dere waltzes and foxtrots—so straight and proper. So goddamed white skinned. Ownin' all dose boats and grass pieces an cattle, an us black folk got nothin'. Beggin' all de time. (Jackson 2003, 42)

While black islander accounts claimed the murders were revenge for being cheated out of a parcel of land by a wealthy white Utilian,[10] Jackson's narration bears the colonial themes of white superiority, with Mcfield angered and jealous of whites owning wealth and performing high society dance and music. In this next part from the novel, Mcfield is narrated as confronting the racist stereotype of the lustful black body, consumed by rage and envy.

> While Olimpia sailed on, his thoughts retrieved other hurts. Though there were some kind acts shown him by many islanders, the pain of rejections blotted them out. One of the worst was the time he was up on the hill looking for mangos behind Miss Samuel's. He was far up in one of her trees searching for the ripe ones. Nobody minded in those days for the trees bore so many fruit they fell by the dozens and rotted on the ground. (Jackson 2003, 43)

For locals, nobody minds these days either when local children come into their yards for mangoes. Twenty-two-year-old Tonia still jokes with a past neighbor of her aunt whose garden she used to raid for mangoes. She and all of her friends would throw sticks at his trees to loosen the delicious fruit, only knocking on his door when she wanted something. Each time I saw the two of them together, he would tease her about how she would come to his door and say, "I need a bag and a drink of water!" Never once did she ever ask for the mangoes she had in tow, but she knew she never had to; a sense of sharing is part of Utilian sensibility and sociality. This holds true even in the narration of Mcfield up in Miss Samuel's mango tree. "Just as he was climbing down with a few tucked into his shirt, he spotted one of the children coming out the back door. It was Carrie Sue, white and pretty, with bare legs obvious in the early morning sunshine. She was near his age, around nine. She had caught the sight of him as he began his climb" (Jackson 2003, 43).

Jackson writes that Carrie Sue asks Mcfield for a mango, which he of course offers to her—being her family's tree and all. Carrie Sue spontaneously thanks him by kissing him "firmly on his brown cheek." The girl's mother sees this and comes rushing out asking him what he thinks he is doing out there. He explains he was just getting some mangoes, to which she responds, in anger, "ManGOS is fine[11] . . . But you keep those black hands off my little girl or I'll have you horsewhipped, you hear!" (Jackson 2003, 43). She orders him to pick up the mangoes and leave. Jackson (2003, 43) writes, "He wanted to abandon the tasty fruit but the presence of her whiteness, size and power commanded him. He picked them up and left, hurt, and confused."

In this passage, Carrie Sue and Robert Mcfield's childhood innocence is abruptly stripped away when Miss Samuel ignites into a racist tirade that teaches the two young friends that Utilian sociality does not extend across racial boundaries. Mcfield is confronted by the "size and power" of white supremacy on the island as Miss Samuel threatens to whip Mcfield, evoking the treatment of his enslaved ancestors. Mcfield doesn't process the multiple meanings of this encounter until years later. As he enters his teenage years, he begins to think further about how it is the whites came to hold so much power on the island. He asks: "Who brung me here, anyhow? How I get to dis place all de way from Africa. Das where dey say we come from—miles and miles away. Far away . . . Why dey gots all de money an we doesn't have any? . . . Well, ol' bee. I gonna find a way to get more" (Jackson 2003, 43–44).

Soon the book shifts to Mcfield's killing spree on the *Olimpia*, where he takes the lives of one person after another and then empties the captain's cabinet of all the money and jewelry he finds on board. Here he encounters a rum flask and takes a drink, which stokes Mcfield's vengeful and lustful side:

> The thought of his father's disgust flashed through his mind. How he became so vexed with him after the old fellow learned that his son drank, and harbored thoughts about colored girls—and white ones too.

He was around seventeen when his father jumped him for behaving crude around girls. He had strong feelings for them—a craving that was constant. Any why couldn't he have them—about the white ones too?

"I hears you been sayin' bad things to white girls," his father began. "It be a sin fer you to be lustin' afta de white ladies—besides, dey could hang you fer foolishness like dat."

"Oh, das how it be den? I mus' leave white girls be, but it be all right to lus' after black ones?" (Jackson 2003, 53)

Mcfield's father invokes the Bible to explain lust in general to be a sin, to which Robert Mcfield pushes back, taking issue with white male power and white men's abuse of black women, referencing the history of the slave trade in the Caribbean and the invention of blackness.[12]

"Here dis, old man. Why be it whiTEY has a right to take black women as dey see fit? And de women be ready to take dem on?"

"It be more so wrong for us to lust afta dere kind."

"Why? Why? Dey be de ones dat make you an me brown . . . I rememba talkin' to dem black ol' timers in Roatán dat de slaves usta be brought right here to Utila to fatten up and be sold in New Orleens." (Jackson 2003, 54)

Jackson's narration both portrays the common stereotype that black men possess an insatiable desire for young, white women and attempts to reconcile (perhaps in the author's mind) why Mcfield would have committed the crime in the first place. In this sense, Jackson speculates that Mcfield was provoked and potentially justified by a history of enslavement and oppression. Throughout the time on the *Olimpia*, Jackson presents Mcfield as confronting this paradox: how he could possess both a fondness for some white woman and an intense rage against white dominance on the island. On several occasions, Mcfield fights to suppress voices and images in his head that would counter his justified rage against the white privilege and power embodied by the passengers on board. Faced with Elsie Morgan, the final passenger, Mcfield loses his focus, veering from his plan to wipe out all passengers. In seeing Elsie, he remembers the kindness of the white girl at the mango tree and then her mother pouncing on him, ruining their moment. His thoughts move to a memory of Elsie. Jackson writes: "More images were arriving, dream-like. He saw the beautiful, white figure on the dance floor that was sprinkled with candle shavings to smoothen it out. She danced by him while he played the accordion, waltzing effortlessly, not missing a step. The white skin of her exposed shoulders, so in contrast with those of his own kind, shining in the glare of the many lighted lamps, had fascinated him at many a dance" (Jackson 2003, 69).

These memories of Elsie and the girl under the mango tree struck a chord with Mcfield and for a moment he considers saving Elsie. He tells her to help him row to shore. But Elsie doesn't trust that he will let her live, as he already murdered all of the other passengers in cold blood, including an infant, tossing the baby and the others into the sea to be eaten by sharks. Elsie takes the oar and strikes Mcfield with all her might in his chest, jumps into the water to begin the long, arduous swim back to Utila. Even if Mcfield would have indeed let her live if she helped him row back to the island, her rebellion had him change course: "Now I *will* have to kill you" (Jackson 2003, 71; emphasis in original). He shoots at her twice, with the second bullet making contact with her arm. Mcfield then panics Elsie with a warning that a shark is behind her. Preferring death by pistol to death by shark, she responds to his call that he'll save her and thus proceeds to swim back to the boat. Once to the boat Elsie is greeted by his gun, and he strikes her with it, sending her underwater. She survives the blow and makes her way under the dory and out of sight. Mcfield searches around and, convinced she's dead, heads back to Utila to quickly gather his belongings and head to the Honduran mainland for an alibi. Jackson's novel follows the journeys of both Elsie and Mcfield back to Utila. For Elsie, she must swim approximately two miles with an injured arm just to make it to the island. Once there, she toils through the harsh island landscape for six days until finally happening upon a coconut garden and being brought safely back to her family in the Eastern Harbour.[13]

Meanwhile, Mcfield lands at Jack's Bight, pushes the dory into the current, and travels by foot to Rock Harbour and across the island via Middle Path toward his home, taking roads Rose (1904, 42) believed to be constructed by the Pech. This back route would reduce the potential for contact with islanders on the way to his home. Jackson narrates Mcfield's journey back in overtly racist and colonial tones, equating Mcfield's blackness with animality and a natural home in the bush.

> A thick canopy of limbs, leaves and vines hung over the path, blocking out the most of the starlight. The strong, tropical bush smell comforted him. He

felt secure and in command in the dark; didn't know why so many were afraid of it. Peoples see duppies in de dark. How come I don' see no ghosts? . . .

. . . He walked easily on, his feet so sensitive to the narrow path that he hardly veered from it. His steps were soft like the padded paws of a prey animal, stealing softly and perfectly, one after the other. (Jackson 2003, 76)

Mcfield makes it back to his home with ease and avoiding encounters with other islanders along the way. Throughout the journey he reminisces about his past and violent sexual encounters with mestizas in La Ceiba (the closest mainland port city) and fantasizes about white women, for whom he has always been denied on Utila. He worries that he will encounter islander boys out hunting iguana but makes his way without anyone laying eyes on him. Once home, he lies to his wife that the blood on his clothing is from hunting alligator. She feeds and bathes him, and he then grabs her crotch forcefully, pins her down, and rapes her. As far as I can tell, there is no historical evidence to substantiate that Mcfield was previously violent against women nor raped his wife, but this fantasy of the black male fits Jackson's colonial storyline.

Meanwhile, Elsie, confronting the same terrain as Mcfield, is lost in its unfamiliarity and hostility. This landscape is threatening, unrelenting, and dangerous. She does not feel at home in the bush the way Mcfield does. Once she manages to swim ashore, Elsie is met with "another frightful hazard directly in front of her, the iron shores, the knife-sharp volcanic rocks barring her to higher ground." There was no away around them and so she had to cross them barefoot, which is like walking on glass and nails. Jackson highlights Elsie's out-of-placeness: "If only her feet were tough like those of the island children who went barefoot much of the time. Being the daughter of Timothy Morgan, the merchant, she was always well-clothed and wore imported shoes" (Jackson 2003, 79). Elsie decides to take a longer route to town, heading westward instead of going straight south, which meant she faced another six to seven miles of treacherous terrain. She didn't know what she was getting into, a young white woman unfamiliar with the bush. Elsie makes her way through the swamp and mangroves, headed in the direction of the Cays: "The view was formidable, for in the distance the mangroves were unmoving, stretching like a stationary army of spider-legged adversaries daring her to advance. She could see no end to them as they swept to the left and then to the right of her . . . All of the swamp was filled with mangrove roots intertwined, forming a natural, hostile barrier" (Jackson 2003, 103 and 112).

Elsie is swarmed by mosquitos and sandflies, suffers from relentless armies of ants, meets an anaconda head on, and spends her nights tortured by thousands upon thousands of blue crabs, which come out with the dark night. For many islanders—especially in generations past—crabbing is an enjoyable activity, with a delicious payoff. Crabs can be caught at night throughout Utila's mangrove swamps, captured easily by hand by grabbing their huge claws and tossing them into a sack. Others will bop them on the head with a small mallet or hammer before grabbing and sacking. Later they will be boiled and seasoned with lime, salt, and pepper for a delicious treat. Yet Elsie's encounter with blue crab turns the hunted into the hunter. Having drifted off the first night on a small stick platform in the swamp, she is awoken by unfamiliar noises:

> Something was coming toward her. These were crabs—thousands of them. They had emerged from their burrows like an awakened and ravenous army to do battle, to spawn—to be out in their time of life.
>
> Elsie peered over her dark surroundings, trying to determine what was happening . . . Something was heading her way. She heard the clashing of their large, fighting claws as they battled for mates and territory . . . Soon they were climbing the stems toward her. Their sharp, spindly legs dug into her long skirt hanging over the stick platform. (Jackson 2003, 114)

Elsie fights the crabs off with club, yelling, "You won't eat *me* alive!" She does this time and time again throughout the night. The crabs never let up, and the oncoming ones begin feasting on those slain before them, creating an "overpowering stench from the dead and crushed crabs all about her" (Jackson 2003, 115).

Jackson's book is a fascinating look into the agency of the more-than-human, and how landscapes and species are entangled with race and notions of belonging. Blue crabs come alive at night—they are out "in their time of life" (Jackson 2003, 114). By day, most crabs avoid the hot sun and high temperatures found on Utila by burrowing and sheltering in the mud within the mangrove swamps. There is a symbiotic relationship between the crab and the mangrove: the mangrove protects and feeds the crab, and the crab contributes to the secondary production of mangroves. Crab burrows are surrounded by the roots of the mangrove, which help protect the crab against predators. As primarily herbivores, they feed on fallen and decaying mangrove leaves. Not only do they help maintain and recycle the energy within the forest by burying and consuming the leaf refuse from the mangrove plants, but their burrows alter the topography of mangrove sediment, aerating the sediment. If the crab were not present, sulfur and ammonium would concentrate in the mangroves, negatively impacting the ability of the vegetation to reproduce (Mangrove Crabs 2021; Pülmanns et al. 2016).

Mangrove ecosystems are important for several reasons. Mangrove forests serve as a buffer zone between land and sea, protect coral reefs and seagrass beds against siltation, absorb pollutants from water and air, and are home to many threatened or endangered species. Mangrove forests also store carbon in their root systems and thus play a role in reducing the impacts of climate change and in combating the effects of rising sea levels, coastal erosion, and flooding.

Jackson was right when he wrote that mangroves can appear as a threatening, impenetrable terrain. They are extremely difficult to traverse for those unfamiliar with navigating rhizomes. For the island's crab species, the mangroves are a safe and comfortable home. The same holds true for Utila's iguana species, which are territorial and claim particular areas of the mangrove forest as their residence. To Elsie, the mangroves form an inhospitable barrier and harbor life-forms that threaten her place within the landscape.

Mangroves are not always narrated as inhospitable barriers. In fact, for Sayda, a young black islander growing up in Sandy Bay in the 1980s, they were a portal to childhood pleasures. Upper Lagoon, the lagoon behind the mangroves that have since been transformed into the Camponado (the shanty development I described earlier that has grown into an expansive neighborhood), was once a swimming destination for island children. Bathing (swimming) in the sea was strictly forbidden on most days by parents, often restricted to weekends and supervised by parents. And so island children would secretly go to this lagoon, far away from the watchful parent eye. Sayda shared that she and her friends would walk a very long way—or at least it felt very long—to climb through the mangroves to get to a swimming hole on the other side of the lagoon. She reminisced, and I paraphrase here: "It was tricky because some of the kids were skinny and other kids were bigger, and it was hard to get through, but we would manage to do it. And we would take our clothes off and go naked so there would be no evidence that we were swimming. This way our parents would not find out because our clothes would not be wet."

Mangroves, thus, can indeed serve as places of refuge and be life giving and life affirming. For the once racially segregated island, it became a place for young people of different skin tones to intermingle without judgement. Mangroves today serve as home to hundreds of new islanders, migrants from the mainland who infill the swamps with refuse and rock to create new beginnings for their families, away from the violence and corruption found on the mainland. Mangroves also create fresh possibilities for "lifestyle migrants" when they remove them along the coastline to build their dream vacation home (see chapter 3 for further discussion).

Ana Deumert (2019) and Laura Ogden (2011) each observe that mangroves stand in place for thinking rhizomatically. In biology, a rhizome is a complex subterranean root system that grows in multiple directions—upward, downward, and horizontally—and new roots can appear at any moment. Examples include turmeric, potatoes, ginger, and of course, mangroves. The rhizome grants a view of landscapes as complex and constantly changing with new relations continuously emerging. Such a perspective is offered in place of the dominant, hegemonic single-origin spatial philosophy of landscapes as emerging because a root system proceeds a plant and then splits into defined outcomes. To think rhizomatically, then, is to acknowledge and capture interconnection and to think beyond linearity. To think rhizomatically is to reckon with multiplicity and assemblage, allowing "one to understand social practices and discourses as endlessly shifting and transforming, as unpredictable, multiple and mobile, creating complex assemblages of meaning" (Deumert 2019, n.p.).

Caribbean scholar Édouard Glissant (1990) argues that rhizomatic thought reflects the essence of the Caribbean experience. As Deumert explains, the Caribbean experience "is an experience of global-local entanglement and relation, in which roots are always multiple." And the mangrove, a tree of the tropics and colonized world, is a special type of rhizomatic plant that enables us to think beyond the Deleuzian rhizome and attend to a history of colonial entanglements (Deumert 2019, n.p.). The Latin word for mangrove, *rhizaphora*, was derived from the Greek words for "root" (*rhiza*) and "to carry" (*phoros*). Translated as "carrying roots," mangroves reveal their roots on the surface. They are visible to us, and they defy binaries of land and sea. They are both marine and terrestrial. The unique habitat of the mangrove denies the very existence of borders and binaries, "collapsing land-bound notions of being-in-the-world, and reminding us not to forget the ocean" (Deumert 2019, n.p.). Nearly all of the pages of *And the Sea Shall Hide Them* are situated within or make reference to the sea, the swamp, the gardens, and life held therein. While the historical novel embraces a one-world-world ontology, reading it through a decolonizing political ontological methodology makes visible coloniality and accumulation by dispossession. One is struck by the many other ways of worlding that might be interpreted and represented by following emergent assemblages and relations to the lifeworld, ones that perhaps include systems of relationality and reciprocity as opposed to the dominant western understanding of the world. Thinking-with-the-mangrove opens up these possibilities.

To "think-with-the-mangrove" (Deumert 2019) in Utila is to notice alternatives to the colonial narratives that dominate Utila's story. To think-with-the-mangrove is to acknowledge that Shelby McNab's account of Robinson Crusoe's time on Utila is his own reckoning with the assembling of his colonial history, as a descendant of one of the first white settler families, with the duppy spirits of his Cayman and the Indigenous Pech, with the bats, turtles, mangroves, and other species that claim space and belonging in this region.

To think-with-the-mangrove would allow us to attend to the notion that mangroves and swamps are not actually inhospitable and unproductive landscapes. Mangroves give life to crabs, iguanas, and juvenile fish and thus in turn to coral, reef fish, seagrass, the list goes on. Mangroves also give life to cultural becomings. As Deumert (2019, n.p.) points out, "mangroves are the place where poor people live and make their living." These spaces give life to new cultural and political movements. We can look to Recife's *movimento mangue* (mangrove movement) as an example of mangroves giving life to cultural becomings.

The *movimento mangue* is a cultural-political movement born in the 1990s in Recife, Brazil, an urban agglomeration built across and on top of several waterways and sheltered by reef along the Atlantic shore. The movement is known for its music, which mixes local and global musical forms, combining Afro-Brazilian traditions with global hip-hop, reggae, and funk (Deumert 2019). Central to the movement and music are questions of justice and equity,

"And the Sea Shall Hide Them" 65

emerging from the history of Recife as the first slave port in the Americas. In the mid-1990s, the movement published a manifesto titled "Caranguejos com Cérebro" (Crabs with brains). The title referred to Recife's mangrove human inhabitants as human-crustaceans. Deumert (2019, n.p.) writes, "Thus with a stroke of the pen, they decentred human exceptionalism. It is a move that echoes contemporary work in post-humanism. The mangrove, in other words, is a place where different ontologies are possible—where humans become part of the natural world, and no longer stand above it."

To think-with-the-mangrove is to see the colonial and racist overtones in *The Sea Shall Hide Them*, as well as to see how humans are part of Utila's mangrove landscape. How could they not be when 70 percent of the island comprises mangrove forest and associated wetlands? For generations Utilians have navigated the waters to get from one side to another and out to sea, and they have learned to trek through them to hunt and eat iguana, crab, and other species. They have built their homes atop the mangroves on high stilts with the smell of the brackish waters part of their everyday sensory landscape.

To think-with-the-mangrove allows one to reckon with the agency of the nonhuman. The sea, for instance, is presented with its own agency in *The Sea Shall Hide Them*. It offered life through the abundance of fish and turtle: "Azure waters were lucrative providers of many kinds of fish, conchs, turtles, whelks and other species" (Jackson 2003, 116). But the sea also took lives, with its ferocious sharks that "can swallow ya whole," and currents and tides that combine with faulty skiffs to swallow lives of many past mariners and fishermen (Jackson 2003, 52). The sea swallows and hides lives ("The Sea Shall Hide Them").

The sea was meant to hide the *Olimpia* massacre. But Elsie Morgan survived and returned to the island to detail the story. The story ends with Robert Mcfield being detained, confessing to the robbery of those on board, the murder of eleven passengers, and the attempted murder of the twelfth passenger, Elsie. Mcfield receives no trial and is taken up Middle Path and hanged from a limb of a mango tree.

Mango trees are widespread in Utila, first arriving in the Caribbean with Spanish explorers of the fifteen century and quickly spreading through the region's fertile soil and warm climate. In many cultures the mango symbolizes love, fertility, and sometimes immortality. It is the national fruit of India, Bangladesh, and Pakistan and is considered food of the gods. On Utila mangoes are plentiful and serve multiple purposes, from comfort food to the "weapons" kids from my generation would use to "play war" back when they were children, to natural medicine that clears out lung infections, combats influenza, and treats tuberculosis. Miss Betty, who listed all the mangoes she uses on the island—"John Bell, Golden Mango, Cutshorts, and the Hairy Mango"—found the Hairy Mango to be the most curative of all: it's all the little hairs holding all the power.

In Jackson's novel, the mango can be read as a symbol of love, lust, and desire but filtered through the lens of Utilian race relations at the turn of the twentieth century. Jackson (2003, 274) narrates the scene at Mcfield's imminent hanging in this way:

> Mcfield glanced at the huge trunk of the mango tree, the fluttering green leaves. A mixture of tempting, yellow ripening fruit and less ripe green globes evoked and [sic] intense gaze from him. Jus' like de ones I picked a long time ago when I was a little boy, he recalled. I comes down from de tree and gives one ta de little white girl. Den she kisses me on the check. Den here comes de big, white woman an' tells me to get my black ass outa dere. Don' be botherin' her little girl. Damn mango tree gonna do me in agin'.

The mango, like the mangrove, is a colonial tree. The story of Elsie Morgan and Robert Mcfield is one that reveals entanglements of race, belonging, lust, love, mortality, conquest, and capital accumulation. Mcfield's ancestors were brought to the Caribbean as part of the slave trade. His family settled on Utila as free blacks but were restricted access to accumulate land and capital in the manner that Elsie Morgan's ancestors did. This history of colonialism, slavery, buccaneering, and settlement has shaped contemporary race relations on the island, even while there is a high degree of intermixture.

In his 1975 dissertation on Utila's remittance economy, Lord (1975, 103) mentions the *Olimpia* massacre as "the paramount example" of how ethnic groups on the island "can act as self-conscious units to demonstrate how clear the ethnic stratification lines are drawn in Utila." Lord notes that in 1975 islanders were still talking about the incident and that while several versions with differing explanations for the piracy, robbery, and murder existed, the outcome is most significant. He writes:

> The end of the affair is most important because when the culprit, a colored Utilian, was caught on the mainland he was brought back to Utila and

promptly lynched. Not only was he hung without the benefit of a trial, but according to one informant was buried in the cemetery in a standing position until public opinion forced reburial in a horizontal plane (even then he was buried on a north-south axis rather than the traditional east-west axis prescribed for Christians in Utila). White outrage over the piracy, even though colored islanders also had been murdered, has derived from assumed colored animosity towards whites (the hangman of the Olimpia murderer had his house mysteriously burn down shortly after the hanging), and subsequently whites have acted as a group to obtain retribution or at the least prevent further episodes of anti-white feeling. (Lord 1975, 103–4)

The stratification that Lord observed in the 1970s continues to some degree today. Lord described a hierarchy in which wealth, power, and the most important positions in terms of local leadership were held almost exclusively by the "so-called white population of Utila," which at the time of Lord's research he estimated to be nearly three-fifths of all islanders. These islanders came to Utila from other parts in the Caribbean where they likely intermixed with people from other ethnic backgrounds, but Lord notes that the white Utilians who were contemporaneous with his study would have refused to accept this intermixture as a possibility, claiming a direct line back to England. These Utilian "whites" all descend from the seven founding settler families mentioned earlier. Fifty years later, this distinction continues to exist, but the vast majority are now intermixed.

Lord distinguished subdivisions within the white population and included a second subdivision of the white population to be those with Spanish surnames. These families—and Lord provides their surnames—married into the founding families and were folded into the white population. They became afforded the same benefits and prestige as other "whites."

Lord (1975, 97–98) places Utilians with African descent into a second tier on a ladder of social prestige, writing:

Second on the ladder of social prestige, etc., are Utilians with Negro ancestry, collectively called "colored." . . . Colored people are not, in general, considered to be mentally or morally inferior to whites, but it was implied by white informants that there was a qualitative difference between themselves and coloreds that would forever separate the two groups even though they lived side by side. At no time . . . did I find the same stereotypes of colored people as I have encountered in the United States (e.g., that they are inherently lazy, immoral and decadent, and the like).

At the very bottom of the social hierarchy—at both the time of Lord's work and my own—are Honduran mestizos, locally referred to by both black and white Utilians as "Spaniards," a negative epithet for individuals with Spanish heritage. "Spaniards" are typically from the mainland, speak Spanish almost exclusively, and work and live in the least desirable occupations and areas. Lord (1975, 98) writes:

> They are typically poor in comparison to Utilians, have to live in the worst housing in the island (due to cost factors and the absolute shortage of rental property), usually have shabby clothing (and little of this), are immoral in the extreme according to Utilians (women have questionable reputations, men and women live in common law union rather than marry according to civil statutes) and epitomize uncouth and uncivilized behavior (e.g., spitting on the floor on one hand, and being satisfied with meals of only beans and rice on the other).

Using the label "Spaniard" hails a particular set of people as distinctly "Other." The use of the identity marker "Spaniard" highlights the animosity descendants of British settler families feel about the surrender to Honduras. To mark someone as Spaniard enlists the colonial history of the Caribbean, the struggle for island territory, and the sense of entitlement early settler families and their descendants feel to the place and cultural logics. To enroll the marker "Spaniard" as opposed to "Honduran" is to deny a common heritage among Utila's contemporary Hondureño population. And to be "Spanish" as opposed to "Honduran" or "Utilian" is to remind a group of their placement at the bottom of a social hierarchy.

The stereotypes of Spaniards that Lord observed in 1970 continue to persist, perpetuated by white and black Utilians and then reproduced by European and American immigrants and tourists. Mainland Honduran immigrants are often the first to be blamed for any crime that besets a tourist. The Camponado, once a swimming hole for island residents, once a home to crabs aplenty, is now written about in tourist blogs as "where foreigners can't go [and] where the drug lords make the laws" (Blog 2016).

◆ ◆ ◆

Like Florida's Everglades described in Ogden's *Swamplife* (2011), Utila's ecosystems are rhizomelike; they are spaces where mangroves claim and rethink boundaries of protection; where the nonnative lionfish has significantly altered the reef ecosystem, leading to kill campaigns by environmental groups; where sea turtles lay their eggs on shorelines protected from local hunters by foreign tourists; and where the habitat of the endangered *swamper* is increasingly clear-cut for housing and tourism projects, ironically all contributing to bring more tourists to the island to engage in "protection" of the very things development is destroying. The assemblages of humans and other species, each claiming territories, but in unique and sometimes contradictory ways, are spaces for forging new socialities, yet to be fully understood or theorized. These are the silences and absences of the contemporary conservation and development model. These elisions are the contradictions of an industry that destroys while claiming to protect. And, these elisions are the focus of the next couple of chapters, which explore tourism development and the conservation narrative surrounding Utila's affect economy.

Right across from the entrance to the Camponado sits the old "Indian Well", a relic from the past that Richard Rose found representative of the industrial nature of the island's original inhabitants. He wrote, "It was a large affair, of circular form, with funnel shaped sides of stone-wall. The people had it filled up many years ago and in its place built an ordinary sized well of square shape and perpendicular sides. It still bears the name Indian Well. And the people have a playful meaningless saying: 'If a visitor drinks 'Indian Well' water he never leaves the island.'" This saying has persisted for more than a century, transforming into the popularized phase "If you come to Utila, you're never gonna leave." The phrase has taken on new meaning among the tourist crowd, as we will see in the next chapter.

Chapter 2

"If You Come to Utila, You Can Do What You Want, and You're Never Gonna Leave"

Utilians are down to earth. Could serve to the very best. They are outspoken and helpful. They will help anyone with a problem, join hands and help fix things. Utilians don't know about fear because they have lived by the freedom of life on this island.

Prior to the tourist industry, Utilians were very reserved and very careful of everything including fruits, vegetables, wildlife, everything. And since the tourism came people have changed a lot. People has changed the style of living and joined in the freestyle of living that the tourists have brought along . . . If you was to look at the local part, the key for us environmental [is] we had a lot more freedom and a lot more choices . . . since now we're more developed there's certain restricted areas that we don't go. I'd say we don't got that freedom like we had before. We'd go sit on the beach and we'd go find a coconut tree. Now there's not much trees or anything left. So, we're restricted from a lot of things. A lot of habits and culture that we had before . . . Those is gone because of the restrictions and the developments. And they cutting the ground and posting private properties.

Taken together, these two quotes excerpted from interviews with women islanders are emblematic of a shared perspective found among many descendants of settler families. In the first instance, one of the earliest hotel operators in Sandy Bay characterizes a Utilian disposition.[1] The quote suggests a close-knit community of locals who care for and help each other out. This disposition emerges from a history as members of a small fishing and agricultural community who, while immersed in the global market, were the very first rung on a long ladder. At the point of extraction, Utilian fishers and farmers relied on volatile markets and at least two additional levels of intermediaries before their products moved outside the region. Thus, families learned resilience and reciprocity, sharing and trading local goods and services. This ethic of care is also closely associated with everyone descending from the same two root families, so most islanders are related in some fashion or another.

The second quote came from a 2016 interview from the team-based research on collaborative conservation. It was in response to a series of questions about local attitudes toward the environment. In this quote, another descendent of one of the settler families who works in a family-owned machine repair and rental shop shares a common sentiment that tourism development has impeded local people's freedom. The speaker's remarks elaborate the myriad impacts of tourism development. The growth in tourism has been accompanied by degradation of the environment, which in turn played a role in the tightening of environmental restrictions. It has also led to increased privatization of land and other environs, including mangrove wetlands, as property values increased. As well, with foreign investment came foreign ways of being, including fencing in property and keeping dogs or other deterrents, such that people did not cross into one's yard. This runs in sharp contrast to the Utila of the past, when islanders could walk with ease through one another's grounds to get where they were going—nobody fenced in. As a now deceased friend once shared, "These are the kinds of changes that restrict Utilian freedom."

◆ ◆ ◆

So, what then, is freedom for Utilians? The way in which Utilians talk about freedom is very similar to that encountered by Melissa Johnson in her study of rural Creole Belizeans. Melissa Johnson often heard the word "freedom" in association with the bush. As in Utila, for rural Belizeans "the bush" refers to the country and the forest, an unrestrictive, undeveloped "nature." For Johnson's (2018, 63) interlocutors, the idea of the bush was tinged with notions of "backwardness" and uncivility, and the idea often took on explicit moral and racist implications The bush was frequently associated with the most raw, undeveloped, uncivilized, uncouth, and unsophisticated ways of being. Yet the bush for rural Belizeans—just as for Utilians—is also associated with very positive meanings, especially as it relates to the idea of "freedom."[2]

In both rural Belize and in Utila, freedom is associated with the ability to move around freely outdoors, as opposed to being confined in a house. It is used locally to refer to a high level of sociality. To be free is to live among and entangled with the natural world, to wander beaches, to sit under coconut and mango trees, to visit with and help one's neighbors, to travel any direction one pleases, unrestricted by private fences, just as one's ancestors had. It also means to be able to work on one's own time, outside of a time-clock orientation of the global capitalist economy. Utilians, like rural Belizeans, appreciate the bush as a place separate from the government, free from its reach.[3] As a result, in both cases, there is a deep resentment for the government putting any kind of restriction on them (Johnson 2018, 64–65).[4]

To be free is to be unconstrained, a concept of freedom that closely articulates with the tourist experience. This chapter explores the manners in which various ideas of freedom have assembled together and how islander notions of freedom and ways of worlding have been challenged through the emergence of Utila as a global destination for backpackers seeking drinking, drugs, and dives. Bolstered by technological advances and an Internet trade in tourist fantasies and images that narrate the island through a particular lens, Utila has been transformed in profound ways over the last two decades, generating socioecological change and giving rise to an emergent multispecies conservation tourism industry. To begin this journey, we start with a brief review of why people travel.

Freedom, Tourism, Discovery, and Dispossession

There is a wealth of literature analyzing tourists' motivations and attitudes, some of which was covered in the book's introduction. In brief, general motivations include a desire for a different experience, apart from the mundane nature of one's daily routine; an opportunity to explore a personal interest in history, culture, or ecology; a desire for thrill, adventure, danger, or sexual excitement associated with "exotic" locales; an opportunity to relax (e.g., on beaches, resorts, or cruise ships); a desire to experience "pristine" natural areas or "untouched wilderness"; a quest for authenticity; or a search for meaning in life (e.g., through unplugging from technology or taking a religious pilgrimage). Many of these motivations assemble together for tourists.

Dean MacCannell (1971) argued that the fundamental motivation for the tourist journey is a universal quest for authenticity, which reflects the human concern for the sacred. MacCannell (1971) believed that all tourists seek authenticity in other times and places away from their everyday lives and are particularly drawn to the "real lives" of others who somehow possess a reality apart from and difficult to discover in one's own daily experience. There is a great deal of romanticizing that takes place in this search for authenticity.

This quest for authenticity is closely tied to modernity, as modern humanity tends to associate modernity with a loss of authenticity. Thus, tourists seek out simplicity, poverty, and purity in other people and places. These characteristics—simplicity, purity, a nonmaterialistic sense of spirituality—are often reflected to tourists in "wilderness" or "undiscovered" environments. Such regions offer the opportunity to experience the sublime, or the awe and reverence that one feels through the power of nature (Bell and Lyall 2002; Mendoza 2018). A sense of discovery thus becomes a core part of what the tourists are after in their journeys to protected areas. These are the types of tourists who land in places such as Utila, tourists on a quest to discover both something about themselves and others (discussed in more detail in chapter 3).

Paige West (2016, 53) found a similar quest for discovery among surf tourists during her research in Papua New Guinea, who were deliberately seeking out "untouched" marine environments. The young men she spoke with were also are in search of "freedom," using the word to describe a place without rules and regulations. For them, Papua New Guinea represented such freedom, and they discursively produced it as constituted by idyllic villages untouched by capitalism (53). Often West's (2016, 60) surfer interlocutors would edit the presence of local actors, describing a new frontier that they have discovered, which has been unspoiled by tourism. This image of an unspoiled frontier articulates closely with Caribbean fantasies.

Postcolonial, decolonial, and feminist analyses reveal how images of the Caribbean reproduce colonial power relations, imagining and narrating lands, seas, bodies, and cultures as open to be occupied, invaded, used, viewed, bought, or consumed (Sheller 2003, 13). The Caribbean paradise is presented through a racialized plantation politic for the Euro-American consumer, and metaphors of paradise and plantation facilitate the reproduction of colonial/postcolonial power relations. These relations are advanced through both explicit advertising that makes use of images of white foreign tourists being served or entertained by happy locals and that position the tourist-to-be as an explorer on the verge of discovery, and through more "subtle practices such as promoting plantations as places of play and relaxation," including the naming of hotels "plantation hotels" (Feldman 2011, 47; Strachan 2002). (Honduras' Cayos Cochinos hosts a resort by the name of "Plantation Beach Resort," and Roatán is home to the "Seagrape Plantation Resort.")

In the material I referenced at the start of this chapter, Melissa Johnson and Paige West encountered similar ideas of freedom in their work but emerging from two different sets of actors: locals and guests.[5] In each case, freedom was defined as freedom from restrictions and structure. Similar notions exist in Utila in that both locals and guests are seeking freedom from rules. For Utilians, like rural Creole Belizeans, there is a desire to be free from government involvement and from regulations and restrictions on their relationship to the natural world. For tourists who come to the island, there is a strong desire to be free from the constraints and norms associated with their home communities and institutions. What tourists are seeking is liminality.

Anthropologists and other scholars have found Victor Turner's (1967, 1969) classic works on ritual process and "rites of passage" helpful for understanding tourism motivations and tourism as part of a liminal process (Graburn 2012). Rites of passage were first described by Arnold van Gennep ([1909] 1960) in his work analyzing the various rites individuals encounter in the course of their lives, such as initiations, marriage, funerals, and so forth. Van Gennep identified three stages in a rite of passage. The first phase, separation, refers to the social and spatial separation from a current status, conventional norms, and social ties. The second phase is the transitional phase, in which the individual is suspended in an "anti-structure," where conventional norms and social ties are interrupted. In the final stage, reintegration, the individual is brought back into the previous social structure, usually at a higher social status.

Victor and Edith Turner (1978) applied van Gennep's rites of passage to theorize religious ritual and to develop the notions of "liminality" and "communitas." The liminal state aligns with van Gennep's transitional stage. It is a time when individuals are "between and betwixt" (Turner 1969, 107), out of time and space, and outside of social conventions (i.e., rules and regulations, both formal and informal). During this phase of the ritual, initiates who partake in the rite of passage together will experience intensive bonding or "communitas," a profound feeling of community with those around you that is associated with the direct experience of the sacred or supernatural. For Turner, communitas develops at the margins of social structure, among outcasts, the poor, or people who have consciously removed themselves from the constraints of society (e.g., extraordinary religious leaders or artists). Some anthropologists borrow from Turner's analysis of ritual process to conceptualize tourism as a sacred journey (e.g., Graburn 2012).

The tourism experience closely resembles a process of sacralization, or rite of passage. Tourists are separated from their normal world by entering a new, unusual setting where they enter a "learner" status, hence a state of liminality. In this unfamiliar space, they are exposed to new and different ways of being—it feels as if "the rules don't apply" because the "rules" are so unfamiliar. When they return home, they are reincorporated into their old communities and structure, but have been somehow changed from their tourist experience, and hence they earn a somewhat higher status among their peers. (I return to the analysis of the status, capital, and distinction that tourists earn and exchange in the market in chapter 3.)

With the suspension of everyday mundane obligations, tourists are given license for permissive and playful behavior. They are encouraged to be unconstrained and to experience a "nonserious" sociality. While they are in this state of liminality, tourists exhibit spontaneity, hedonism, and a sense of freedom from structure and everyday life. Their social reality has been inverted in that the tourist feels at the top of the social rung while in the tourist setting. Several of these traits are masculine in nature: hedonism, a focus on the self, a feeling of superiority, and license to be dominant in a foreign place, and so on.

These narratives of discovery feminize the landscapes and peoples of the Caribbean, positioning them as open and "readily available for imperial penetration" (Feldman 2011, 47). Despite the centuries of landscape manipulation and ecological devastation, the Caribbean continues to be commonly presented as empty, unoccupied, and ripe for discovery (Bolles 1992; Feldman 2011; Strachan 2002). Tourism advertising pitches Caribbean destinations as "unspoiled," with local populations presented as free of conflict, politics, and behavioral restrictions. We can see evidence that tourists believe they have found this authentic place where the "rules do not apply" in Utila.

The AboutUtila.com website resembles most travel guide depictions of Utila. The site, which former mayor Alton Cooper approved to give the tagline "the official website of Utila," starts its "about" page in the following way:

> Welcome to our beautiful *tropical island of Utila*, nestling in the Caribbean Sea . . . and surrounded by vast coral reefs with prolific undersea life. The smallest of the major islands in The Bay Islands group, we are renowned as being one of the *least expensive* and most beautiful places in the world for a *scuba diving vacation*. Whether you are on a back-packing tour to travel Central America or just on a short Caribbean vacation away from the busy office, the friendly island people will ensure your time spent here will be a relaxing and delightful experience you will treasure and want to share with your family and friends. From the Payan Indians, through the years spent as a British Colony, to the present day as part of Honduras, our island provides a rich and diverse cultural experience with it's [*sic*] unique blend of British, American and Spanish *heritages* making this an authentic Caribbean Island which, even today, is still a largely undiscovered and unspoiled *tropical island paradise*. (AboutUtila 2021; emphasis in original)

The host of this website is an Australian who has been living on the island since 2003 and who works in island real estate. That this website is the top Google hit and continues to be presented as the island's "official website" says something about how official narratives of place are created and presented to the broader world.

More and more, Utila's image is being crafted by foreign-born island residents who have captured and cornered a large percentage of the local real estate and scuba tourism industry. Utila is presented in most online features as a laid-back, Caribbean paradise filled with locals without a care in the world, living happily, ready to receive and host tourists. This narration of islanders dispossesses local people both rhetorically and physically. Locals are rhetorically dispossessed of their experiences and knowledge of the local environment. The presentation of Utila is filtered through the dive tourist's lens and obsession with undersea life rather than the local's intimate understanding of their landscape. The latter would instead underscore a fisher's ability to find where the groupers are running in the pitch black of the night more than forty miles out, or a local's knowledge that blue crabs are most plentiful at dusk off the path on the way to Pumpkin Hill.

Utila has experienced dispossession two times over. First, the Pech were decimated. And, they remain altogether disappeared by the absence of their own identity label of "Pech" (not "Paya"). All tourism materials (including the above website) still refer to the island's Indigenous population as "Paya," even though this is a colonial label not embraced by contemporary Pech (see figure 8). For example, Roatán publishes *Päyä: The Roatan Lifestyle Magazine*, a bimonthly English language magazine "about the essence of what is Roatán: its people, history, nature, and traditions" ("What Is Paya Magazine" 2020). The online search queue of back issues does not bring up a single mention of the "Pech," though there is an article by the magazine's British history editor called "Paya Resistance" which recounts the Pech's ("Paya's") resistance to Spanish slavers (Tompson 2019).

The Pech were honored and glorified by the island's next wave of settlers (today's Old Heads) after their disappearance—as one can see in the writings of Richard Rose, William Jackson, and Shelby McNab that were presented in chapter 1. The same is true when you speak to descendants of the Old Heads; they revere the first inhabitants, knowing all too well how difficult it was in the early days to live off these lands. While the Pech are celebrated by current inhabitants, both descendants of early settlers and newcomers alike, the continued use of a colonial label rhetorically dispossesses them of past and future claims to the island and Cays.

Figure 8 Absurd cover choice for map of Utila island published by a magazine named after the colonial name for Pech, *Päyä: The Roatan Lifestyle Magazine*. Photo of map cover by author.

I said Utila's peoples have been dispossessed twice over. The second form of dispossession is contemporaneous with the writing of this book. The knowledge islanders possess of the multiple ecosystems and their inhabitants are overlaid in tourist and housing markets with empty shores and beautiful waters awaiting foreign desires. This rhetorical dispossession has material consequences. Materially, locals are increasingly losing their control over land and industry. The two are tied together—material dispossession happens through rhetorical dispossession. First, you remove the people and their links to a place. Then, you can move in—seemingly justifiably, as the place is "empty"—and take ownership of a place and its future. Home and Garden Television (HGTV) has helped this unfold in Utila. HGTV's *House Hunters International Caribbean Life* has twice featured homes on the island, once in 2014 and once in 2015.

Both HGTV episodes feature an erasure of Utila's past and present and are not factually representative. For instance, in season 4, episode 14, "Lawyer Trades City Life for Relaxing Beaches in Utila," Sandra, the woman we follow on her quest to find the perfect home, tells us that she has traveled extensively through the Caribbean, but no place fit her as well as Utila. She loves Utila because "it's laid back, a little bit of buzz, and [has] beautiful pristine reefs, and the people are friendly and nice." As Sandra tells us how wonderful the people are, the camera zooms in on two brown people, both smiling widely. The first features a man carving coconuts and the second is a woman with a monkey on her shoulder. It is questionable if either of these people are actually from Utila, as there are no monkeys on the island.[6]

The second episode, "Half Price Paradise Utila Honduras," which aired in January of 2015, follows a couple relocating to Utila from Chicago. The husband runs an online business and can operate it from anywhere provided he has good Wi-Fi, and his wife plans to begin her own massage business on the island. Like Sandra, they both love to scuba dive. Throughout this episode, the HGTV narrator reminds us how "cheap" Utila is. They compare prices in the show's trivia notes for the costs of everyday items and land in Utila versus Chicago. The couple has a budget that enables them to look at homes up to the $360,000 mark, a price range that will include a fully furnished home plus furnished rental units on the property. This price is far too high for a local islander to afford in Utila's cash housing market. The couple is blown away that a place at that cost comes with so much square footage, includes "very modern," "American"-style appliances, and will require them to only pay $125 in annual taxes, whereas in Chicago they were paying $11,000 in taxes. Walking through this $349,000 Tradewinds house they remark "this would be a million-dollar house in the suburbs (of Chicago) and you wouldn't have these views!" "Your money goes further in Utila," the Realtor tells them in response to the couple's delight and surprise at the low cost. The postpurchase follow-up scene shows the couple living happily on the island, running two businesses, and "saving nearly $70,000 a year" by living in Utila rather than Chicago.

❖ ❖ ❖

In sum, with the rise of the Internet in the last twenty years, the "idea" of Utila is now distributed from past travelers to future visitors, snowbirds, lifestyle migrants, and retirees, which is having tangible impacts on the local population and ecology. The "idea" of Utila and the search for freedom and a place without rules has also led to an uptick in hedonistic tourist behavior, causing increases in alcohol and drug consumption among younger generations of Utilians. As the speaker from this chapter's opening quote says, "People has changed the style of living and joined in the freestyle living that the tourists have brought along."

Enter Dr. John, "He's the Doctor with no T-shirt on."

In 2016, I sit with Marta, a Garifuna woman, and Antonio, an Italian conservation scientist. Marta shares: "Many people think the only thing to do on Utila is to dive, but there are *so many other things* to do on the island." Unfortunately, she says, "what is being sold to the world is dive, dive, dive." "And party," Antonio adds. Marta agrees: "It is Dr. John. Sunjam. Rehab. Carnival." All bad looks for the island, from Marta's tone of voice. Dr. John is a notorious American doctor who moved to the island in 2002 and has taken on a celebrity status by engaging in and encouraging debauchery among tourists. "Sunjam" is an annual overnight rave on one of the private Cays, which began in the late 1990s alongside the rise of dive tourism. Rehab is a rowdy bar over the water. And, Carnival, a local tradition in all Honduran towns, has morphed into a weekend-long binge that attracts high numbers of national, regional, and foreign tourists to party on the island.[7]

After her list of unvirtuous people, places, and events for which her home has become known, Marta shares, "When I saw 'THAT video' and the

Figure 9 Screenshot of Dr. John in his house posted in John DuPuis's online article "A Visit to Famous Dr. John in Utila," *Honduras Travel*, 2017. Retrieved October 19, 2020 from https://hondurastravel.com /featured/visit-famous-dr-john-utila/.

number of likes and views on it, I wondered, who is going to pay for all the barrels of garbage that people are bringing here now, as a result of that video? From all those tourists who are coming to party here?" Marta is referring to a YouTube music video gone viral that helped make Dr. John famous, with the unfortunate chorus line of "If you come to Utila, you can do what you want" (MartyandGinski 2013). The video, made by two Australian tourists who spent a year on the island, showcases a misogynistic and party-obsessed island lifestyle, filmed during Carnival season. The YouTube description under the video encourages people to visit the island: "Utila is a small island off the coast of Honduras. It's the craziest, most fun, most life changing place you'll ever visit, and this song captures much of that experience. Enjoy the song, share the song, sing the song and visit Utila."

The video had accumulated just shy of a half a million views by August 2020, seven years after its original posting. Many islanders were embarrassed by it. It popularized "Dr. John" (figure 9) and transformed a fairly recently arrived (under twenty years) American resident into a "must-do" for tourists making plans on the island. John Dupuis (2017),[8] creator of *Honduras Travel*, an online travel blog that is the newest in Dupuis's travel tip publications, writes:

MEET FAMOUS DR. JOHN IN UTILA!

It was a perfect match for Dr. John and Utila. For Utila, a full time resident doctor was now on the island. For Dr. John, he was a member of a tightly knit community on a small Caribbean Island. Best of all, he could enjoy a good party after hours meeting all the foreign travelers that passed through the island. Perhaps his biggest life changing event was when he treated an Aussie tourist who was visiting the island. The traveler was in Utila taking a series of diving courses when he became ill and visited Dr. John. That was the beginning of a friendship that would cast him into fame. You see, Marty, and his friend Ginsky spent a year in Utila. They had such a blast, they decided to share their experience by producing a video. The video, titled "If you come to Utila" was a grand success and went viral as soon as it was uploaded to UTube. The video features "Dr. John" a doctor with no shirt on, who is nearsighted and likes to party. It's hard to believe, but this describes Dr. John quite well!

John Dupuis's 2017 article in *Honduras Travel* details his visit to Dr. John's house, an excursion many a tourist will take, often more than once. Dr. John's house is located just on the outskirts of Utila Town, close to BICA's headquarters. John Dupuis writes, "If you go to Utila and do not visit Dr. John, you cannot say that you [have] been in Utila." (I have been to Utila countless times over the past twenty years and have never visited Dr. John.) According to Dupuis, you can usually find Dr. John on the porch of his bright pink house between noon and 10 p.m., resting up before he goes out at night. We are warned: "Keep in mind that he is a party animal. This means that he parties through the night, so he is not a morning person . . . [But] If you like to party, your chances of running into World Famous Dr. John at one of local bars are actually quite good."

From what I gather, a visit to Dr. John includes participating in his drinking challenge, after which you are rewarded a Dr. John T-shirt and a photo with him. The drinking challenge involves listening to AC/DC's song "TNT" and taking a shot of hard alcohol mixed with powdered juice each time "TNT" is sung in the chorus. I believe that makes for four hefty shots in fewer than three minutes. This is how Dr. John makes his living on the island (through payments for the drinking challenge and T-shirts). His latest popularized T-shirt features his moniker "La Mera Verga," something which a tourist from Tegucigalpa called him and which Dr. John was so flattered by that he adopted and branded himself with the compliment. Verga is slang for penis in Spanish and the phrase "la mera verga" is used in calling someone "the shit" or "the fucking best."

While the drinking challenge is how Dr. John currently makes money, he apparently moved to Utila from Akron, Ohio, in 2002, after visiting the island for a monthlong vacation and seeing that the island did not have a resident doctor. It seems he did practice medicine for some time, and some have said he was a very solid physician. After the YouTube video, Dr. John's social media presence hit the roof. As of July 2020, he had 17,700 Instagram followers.

The YouTube Video

The video (figure 10) begins with an adventurous looking twenty-something-year-old female hipster walking into a travel agency named Peterpans Adventure Travel. Here she seeks advice on where to take her next vacation. The travel agent asks her a series of questions and learns that this tourist "always wanted to go to the Caribbean" and "would love to try scuba diving." The agent responds, "Cool. Good. Do you also want to party?" She shakes her head affirmatively and gestures with a smile that she is about to let her in on the perfect location. "Alright, good. It sounds like you want to go to Utila, an island in Honduras," she smiles. Upon mention of "Utila," a disheveled, dirty, scruffy-haired man with missing teeth around the same age as the tourist appears in the office and warns the young woman with great concern, "Don't go to Utila!" (It's Marty, one of the filmmakers.) "Why not?" she wonders. He retorts, "Oh, you can go to Utila! But you will never leave!" The travel agent whispers to the young woman planning her trip that "legend has it, he went to Utila for three days and stayed for twenty years." The YouTube clip then launches into a music video in which Marty gets off the boat, dressed squeaky clean, until he is corrupted by the drinking and party culture of the dive industry. The opening lines of the song are

> *If you come to Utila,*
> *you will dive.*
> *But you never ever feel more alive.*
> *If you come to Utila,*
> *you're gonna drink tequila.*
> *If you come to Utila,*
> *you will dive.*

Then the main character (Marty) starts his narrative rap about his experience getting Professional Association of Diving Instructors (PADI) certified. After studying PADI videos and guides, he says he was ready to attempt div-

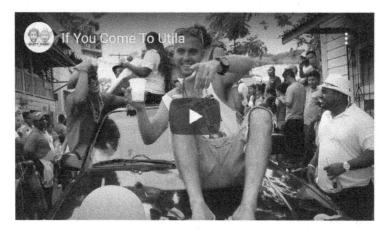

Figure 10 Screenshot of *If You Come to Utila* posted in John DuPuis's online article "A Visit to Famous Dr. John in Utila," *Honduras Travel*, 2017. Retrieved October 19, 2020 from https://hondurastravel.com/featured/visit-famous-dr-john-utila/.

ing but became distracted by "Tequila Tuesdays." On his walk into Tranquila Bar, an open-air bar over the water on the bay next to Parrots Dive Shop, the man is warned by other tourists to be careful. He doesn't heed the warning, and drinks so much that he is sick by the morning. Off he goes to find Dr. John so "Dr. John can make him right." As the video transitions to the chorus, more tequila appears, now being poured over the body of a white woman in a black bikini, from her breasts to between her legs. A group of all white young people sing the first round of the chorus:

If you come to Utila.
Where the water's clearer.
We will drink tequila.
We will spend lempiras.

Then the main character sings back, with his arms around five women in bikinis,

But you won't ever want to leave.
You never want to leave.
In Utila, you can do as you please.

The group agrees, singing, "You can do whatever you want, but you'll never ever leave." This section is overlain with footage of two young women in bikinis showering together outdoors and tourists partying during the Utila Carnival. It ends with a white woman sticking her tongue out at the camera.

The next section of the video is presumably the story of another white male who decides to extend his stay on the island, saying, "I'll just stay another week until I get my Skid Row T-shirt." Eight young women, all wearing Skid Row T-shirts, surround him singing and dancing about how they now have to do shots in order to earn a T-shirt.

We follow the man into Skid Row bar with the young women continuing to sing: "You have to do shots. Lots and lots of shots." They toast him as he works on his five shots.

Each time the chorus is sung, there is more footage of tourists partying, including young men licking tequila off women's bodies, young women showering together, and various scenes of tourists doing "whatever they please." It goes like this:

Lyrics	Scene
If you come to Utila.	Young people drinking and partying during Carnival.
Where the water is clearer.	Footage of a shark swimming with its head stuck in a bucket.
You will drink tequila.	Young white man licking tequila off a young white woman's body, whose face does not appear in the shot.
You will spend lempiras.	Another young white woman laying on her back in a bikini, with lempiras being thrown all over her body.

We see Ginsky (the other video maker) as a dive instructor riding his bicycle off the pier into the water and then transition to him rapping about what the divers may see when out on their dive. In this part of the video Ginsky does a series of silly movements, invading the personal space of those in the shot, as he describes electric eels, lionfish kills, and stingrays. With the latter, he appears to pinch a blond young woman on her upper thigh or between her legs (I suppose illustrating a sting). We hear about the "3 lies of Utila": (1) "I love you." "No you don't." (2) "I'm leaving tomorrow." "No you won't."

(3) "I'm not drinking tonight." "Yes you will." "Because the facts are the facts, you are in Utila still."

While the majority of the 440 comments on YouTube were enthusiastic endorsements of the tone and content of the video, a handful of commentators expressed grief over the transformation of the island through the rise of dive tourism and associated partying.[9] For example:

> AYAHUASCAYAGE, 4 YEARS AGO
> Having known Utila since the 70's, it's sad to see how utterly ruined it has become. It's just like Kuta and Legian, another tourist cesspool of fun, depravity and hedonistic adventure. I remember Elvis, Mr. Thompson, Jackie Cooper and Crazy Fred the Diver and the old Bucket of Blood. I rented a house near the well on Blueberry Hill and paid $3 a night. Except near the few bars, the town was quiet in the night, and all the people were locals. Poor Utila doesn't seem so wonderful anymore.

> LENNY ZAROFF, 3 YEARS AGO
> TRASHY PEOPLE only trash the places where they go. mainly the USA trash. They are NOT welcome in Honduras.

> DAVE NELSON, 1 YEAR AGO
> In 2000 when I was there it was perfect. Now it looks totally Cancuned out. Full of tuk tuks and drunkards. No me gusta mas!

These commenters are nostalgic for a past when the only hotels on the island were small family-owned boardinghouses (i.e., Blueberry Hill was one of the first on the island) and tourists were not kept confined within dive shops. Money used to be spread around across families, from Mr. Thompson's bakery, to the old bar "Bucket of Blood" up on the hill. These earlier tourists also seem to value the nightlife and diving scene, but they imagine Utila at a smaller scale, in the years before 24/7 electricity and, thus, a much quieter and less densely populated locale.

In a few instances, some newer residents on the island pushed back on the critique, and a public debate ensued:

> BEAUTIFULDEVIOT, 1 YEAR AGO
> These people bring MONEY to the island. Tourism is the main source of income for Utila. You might not like that they are on holiday and enjoy-

ing themselves, but their money is what has lifted the island and created jobs and opportunities for locals. What would be sad is if no tourists ever discovered Utila and Utila hadn't developed their tourism niche. You are seriously stupid if you think Utila would be better off without tourism. AND it's the best place on earth. Quit moaning and go home

REPLY

BEN ANSON, 1 YEAR AGO

Are you even Honduran? Mate why don't you fuck off home? I just realised that it's not worth responding to your lame arguments because it is all money-centred. It's people like you who are to blame for the ruin of many beautiful places on earth, islands such as Utila and Ibiza, places in Greece etc. Money, money, money. Get stuffed.

BEAUTIFULDEVIOT, 1 YEAR AGO

I'm married to an Utilian and yes I live here. If it weren't for tourism, none of us would have jobs or businesses. Keep your negative comments to yourself. This video is a light-hearted tourism promotion video and it's doing it's [sic] job well. Utila is nothing like Ibiza. It's still cheap, under-developed and not at all commericalized [sic] compared to Roatan next door, let alone any of the other places the commenters above have compared it too (Kuta and Legian . . . REALLY!!—they are NOTHING like Utila). This video is taken on Carnival, which is a proud party for all Utilians. Go stay on the cays by yourself if you don't like to be around people enjoying themselves.

In the above, Ben Anson compares Utila to global beach hotspots such as Ibiza, Spain, in the Mediterranean Sea, and Kuta and Legian, on the coast of Bali in Indonesia, places that have become known—and criticized—for their wild nightlife. Beautifuldeviot, who presumably also first landed in Utila as a tourist but later married a Utilian, argues that Utila has "developed their tourism niche" previously *because* tourists "*discovered* Utila." Beautifuldeviot's understanding of Utila as a place of paradise waiting to be explored, occupied, and "lifted" up through the creation of jobs for locals,[10] to serve playful and "light-hearted" tourists, fulfills the Caribbean fantasy and reproduces colonial power relations.

In 2016, I asked Marta, the Garifuna woman I mentioned earlier, how people feel about the video and she said:

A lot of people don't like that because that is not the island. You have church going people here and that doesn't appear there anywhere. The way Utila is sold is as if it's all about boozing. And the worst part of it [the video] is that line: "if you can come to Utila, you can do whatever you want." That is what the video says. And you know what they mean when they say *whatever* [*she emphasizes the word*]. You know the meaning of "*whatever*"; whatever comes to mind happens. It's happening. *Whatever* happens there at Sunjam.

Marta is especially animated when she mentions Sunjam, the annual all-night rave that occurs in August on Water Cay. We were just a week away from the 2016 party, a time when tourists from all over the world arrive to drink, trip on hallucinogenic drugs, take ecstasy, and dance. I've never been, though I have been on Utila during the rave—it's quite peaceful! Due to my detachment from this part of Utila's tourism scene, my data are limited to online accounts and interviews with those who have attended. The following "survival guide" from a New Yorker's blog *Alex in Wanderland* (Baackes 2011) gives you a sense of what this annual party entails:

SUNJAM, AUGUST 2010
I'm breaking my normally chronological posting in honor of next weekend's big event in Utila, Honduras. Last year I attended Sunjam Utila with four of my favorite people, ever, and it's one of my favorite travel memories, ever. This post is long but its [*sic*] all information that I searched long and hard for before attending and mostly came up short, so I hope this becomes a resource for future sunjammers. I plan to attend more wild parties like this around the world, and maybe start a Survival Series!

What: Ah, Sunjam. A teeny tiny blip of sand and palm trees in the middle of the Caribbean Sea; normally deserted, but for 24 wild hours each year, heaving with techno dancing backpackers from far and wide.
    There is something thrilling about knowing you are so far from civilization, about meeting people from all over the world in one small stretch of sand, about being able to dip into the warm ocean water in between sets, and about watching the sun come up having never fallen asleep.
    Sunjam began 13 years ago when a group of travelers and friends wanted to have a special party in beautiful surroundings, similar to the origins of Thailand's infamous Full Moon Party. It has gone on to attract famous DJs and make a serious mark on the international party circuit.

The next section of Alex's blog emphasizes the sheer numbers of people (two thousand global backpackers and Hondurans) crammed onto a small private Cay to dance for twenty-four hours straight. Alex describes the potential danger these "true party animals" face in a section she labels "Staying Alive." Here the reader learns of the ready access of drugs throughout the rave and that partygoers should bear in mind that they are on a "deserted isle off a small island off of what many consider a third world country." She recalls seeing a girl go into a seizure and quickly be taken away by boat. Alex appears surprised by how, "miraculously," everyone seemed to know just what to do. This is perhaps surprising because of the "third world" (her words) status of her location. She ends the section with her "most important" advice: "Go Wild." Then, in closing out the blog we return to Alex's routine and mundane existence:

> On a normal night, you're just as likely to find me watching Law & Order reruns as out dancing on tabletops. But I knew this was a once in a lifetime event, so I made myself enjoy every moment, from sipping Johnny Walker on a wealthy Honduran's mini-yacht to sitting covered in sand and sweat and cheering for the approaching sun. If you can possibly get to Utila this week, do it. You won't regret a minute.

As you can see, Alex, the blogger, suspends her normal routine and rules to partake in this wild party on an "uninhabited" Cay in a "third world country." The place is a normally deserted, nearly undiscovered, paradise that somehow almost two thousand tourists find every year to sip Johnny Walker and dance to techno music.

Marta is not alone in her concern for the image that Sunjam creates, and some Utilians that do not partake in the party culture worry that this presentation of the island has contributed to a decline in respect for Utila's women and created a life sentence for young islanders to be forevermore confined to low-paying service jobs associated with the party scene.

Men Who Hate Women

I have come to say goodbye to Miss Ellie and Nilda, the mestiza woman from the mainland who cares for her. Miss Ellie is one of the oldest women left on the island, ninety-four and still going strong. She married into one of the early white settler families, moving from Roatán to Utila as a young woman. I find Miss Ellie and Nilda that morning resting on the porch, along with Estela, a younger relative of Miss Ellie. They were holding a memorial pamphlet for an older woman whose house burned down the night before and took her life. They asked if I had heard the news. I had. It was all over town. This was my third goodbye visit of the day, and the third time I heard the story. This was no accidental fire. It was arson by the woman's son. Locals were distraught, as the woman who passed away was one of the kindest souls on the island, always with her door open to lend a hand or ear. I never met this woman, nor her son, but according to Nilda the son "was a *really* bad man" who "hates women." Upon saying this, Nilda turned and motioned to Estela and said, "He hurt her too." Estela shook her head sadly. I didn't press these details, as I had heard what happened to Estela from another person earlier in the day. The son was known to have physically abused several island women, raping at least two others. A third woman was found tied up in her house, narrowly escaping rape because her friend dropped by her house and scared the rapist off. The women associate these crimes with a rise in male violence, behavior brought on by widespread drug consumption and destabilization coinciding with the growing tourism industry.

Earlier that afternoon my daughter Amalie and I were at Mr. Austin's home, saying our goodbyes. Mr. Austin, phenotypically brown but someone who identifies as white, was in his mid-seventies and, like Miss Ellie, had seen the island changing rapidly in front of his eyes with tourism development. With tourism development came a feeling of heightened insecurity among Utilians, associated with the population outpacing the ability to recognize new faces on the island, rising drug and alcohol consumption, increased petty crime, and accusations of sexual assault. Mr. Austin tells us how he always tries to keep tourists safe, most notably from being raped. He discloses that there are reports of a tuc tuc (a motorized rickshaw) driver taking girls up to Pumpkin Hill and raping them. This was not the first day I heard that women have been assaulted on the island, and two of my conservation friends were jumped, beaten, and mugged on their way home in the middle of the night in 2017.[11] While crimes of opportunity and petty theft have increased with the rising population, I still often wondered when elders talked to me about their fears that sexual assault was on the rise was just that: fear—not imminent and increasing danger—brought on by lack of familiarity with new faces on the island. I still wonder how much of islander worries are caused by the changing demographics on the island.

These were the warnings one hears in talking with local people. They are new warnings and new crimes, according to locals. In other words, this is not the Utila of the past. In the past, the community protected each another and cared for the tourists. In the past, Utila was a small place and everyone knew each another. They also knew the tourists, who came in handfuls not boatloads. But Utila today has been transformed through the growth of tourism and associated labor migration. Its growth has not come without problems, and, according to some locals, it may have given license to sexual predators on the island. Many an elder woman would point directly to the first dive shop as the root cause of decline in respect for island women. Utilians were—and many (especially adults over fifty) continue to be—religiously conservative and prefer modest clothing and abstention from cursing, drinking, and drug use. When I first traveled to Utila in the early 2000s there were still restaurants that refused entry to shirtless men or women in bathing suits. In fact, when the first dive shops started, shirtless men were put in jail. Miss Mary, a white female descendant of one of the earliest settler families and a former chief of police, shared:

> When the divers started that's when the dope started to come in . . . They would go around without shirts. I was the first person to start enforcing the law in Utila . . . You couldn't go around without shirts. You couldn't use these shorts—Miss Nelly [a pseudonym] had a little store and when they put up their leg, everything they owned would fall out. When she reported to me—she had three daughters—she said, "you got to do something about this guy." I took him down to the office and he said, "I didn't know." I said, "well you don't have no pride in yourself?" "Yes, ma'am" (he cowered). I said, "well don't you do it again!" He musta forgot or thought "that old lady ain't gonna do anything." He had a surprise! I threw him in jail for three days with the mosquitos and sandflies!

This former chief of police continued to enforce the local laws, locking up any diver who broke the island's dress code. Soon one of the dive shop owners came up to her house and offered to "make a deal," saying they would pay her $800 a month in order for her to turn her eye away from the divers. She told him, forcibly, to get out of her yard, "You don't buy Honduran law!

And because I'm a woman, you think you can buy me with US$800 son?! No. Please go. Don't you be coming back here if you be coming back with that. If you come back and you woulds like to make a deal that we talks to the divers there, let them to respect the ladies of Utila, and *especially* children. Never seen him again in my yard."

Ms. Mary continued to run the town by the local law, even under threat by the divers that the American consulate would get involved and make it impossible for her to ever get a visa to travel to the United States.[12] As the years passed, and this police chief moved into retirement, the number of divers soon overwhelmed the town, and enforcing local clothing codes became a minor offense that was likely not worth the effort once overshadowed by the rising drug culture and associated petty theft. The dive tourist's attire (or lack thereof)—for some locals—was the start of a slippery slope into a long line of illicit behavior. Tourism is seen in the eyes of some locals to be sending the wrong signal about the island to the broader world. It is sending the message that once on this island, "you can do whatever you want." Marta often reminded me while a tourist binging and partying on Utila likely isn't acting that way when they are home, they model behavior for young islanders, who after high school have extremely limited options on the island. Marta observed that tourists think: "It's okay for me to act this this. I'm on vacation." But, she noted, tourists do not realize that the young islanders they are having fun with in the bars during their two-week vacation are in those bars fifty-two weeks a year. And that while it was fun for a year, or maybe even two years, what happens to these young people after ten years of that lifestyle?[13]

Volunteer Tourism and the Party Scene

This is a book about volunteer conservation tourism, and so you may be wondering why I am sharing so much about the dive tourism industry and its reputation. How do volunteer tourists depart from the typical dive tourist? Do volunteer tourists engage in similar degrees of partying? To answer this question, let's take a look at the field notes of a research assistant on our team in 2016.

In 2016, one of the master's candidates who joined Suzanne Kent and me for our joint project on collaborative conservation rotated through conservation organization volunteer programs, spending one to two weeks in each of our partner organizations to learn about the programs and volunteer motivations. The student shared her field notes with our team as part of the larger study (Field Notes from Utila 2016).

Of the twenty segments that Suzanne and I coded for "partying" (code used to refer to the party lifestyle on the island), fourteen were for experiences coded within the student's field notes. The six others were references interviewees made to the island's party reputation, always in a negative light. I present excerpts from this graduate student's first two weeks of field notes. Each new entry reflects another day.

WEEK 1 FIELD NOTES:

(1) Walking back with [the organization's director], ran into volunteers from station who were going out for the night, so we joined them at The Venue (next to public beach).

(2) ... Played card games in the kitchen then went to carnivalito near BICA at the last minute ... We walked around for a moment then went into Jade Seahorse and sat on deck where we could still see the crowd. The streets were very packed on the way over—looked like mostly local people. There were really large speakers and a bar area set up at the intersection, along with street vendors selling food and a few carnival games set up in the street. We went up to Treetonic, the bar over Jade Seahorse, and it was totally packed with foreigners. Where we were sitting, though, we could still hear all the music from the carnivalito and the crowds below.

(3) ... Came back to station—[Names two volunteers] feeling sick so they took an early night. [Names of two other volunteers] go out for Ladies Night at Tranquilas—free drinks for ladies from 9–12.

(4) We had some coffee and ate and went down to the beach around 6:30, and even though there was still a truck there playing music the party was already mostly over. We walked over to Tranquilas (dead) then moved to La Cueva where there were both locals and foreigners. Halfway through our time there a live band came on the patio to play and people came up to watch and listen from the street. We went back to Tranquilas which was by then totally crowded with foreigners/dive shop people.

(5) ... Went to Tranquila's with [names of volunteers from Kanahau] ... since it is a 45-minute walk into town [from Kanahau's research sta-

tion] they generally only go out on nights with drink specials—Tequila Tuesday (free shots of Tequila if you know what the "code" is—tonight is if you are wearing black) and Ladies' Night (free shots for girls from 9–11 or midnight) . . . The Iguana Station volunteers came over from La Cueva and sat with the Kanahau volunteers for a moment but then we all still ended up standing inside. Bar was filled with foreigners—only a few locals there earlier in the night but they left as the place became more crowded. Talked to four American guys who were on Utila just for diving/holiday while they were traveling, two are leaving in the morning.

(6) . . . Got to Skid Row (all foreigners there—mostly American when we got there though others from a dive shop came in while we were there) and [names a female volunteer] was the only one who did the Skid Row challenge (involves taking 4 shots of guifitty[14]—local herbal medicinal rum).

(7) . . . The three girls and I all went to La Cueva and then to Tranquila together. I've noticed now that La Cueva is basically staffed by Parrots Dive Shop employees—one other time I was there was a Spanish-speaking girl behind the bar, but otherwise the boys at that shop seem to rotate going behind the bar. Tranquila was also almost totally filled with divers from Parrots and other foreigners/tourists.

(8) . . . [someone from Kanahau] will be leaving for Roatán in the morning so she will be staying in town and invited anyone who wanted to stay in town after drinking tonight to stay with her. Vinyl and Tranquila (who are owned by the same person) are offering a joint drinking challenge tonight so most volunteers plan on going in town for that.

(9) . . . All volunteers except [one male] went out to La Cueva—we ran into [two female volunteers] on the way and then [two other women volunteers] met us there. Again, mostly people from Parrots behind the bar and visiting it, but when we went to Tranquila it was mostly locals (save a few people from Parrots). This was the first night I'd seen more Hondurans out than foreigners but I'm not sure why they were all there that night (no local DJ or anything around).

(10) [Two volunteers] said the Iguana Station volunteers went out last night—the Dive Masters from Parrots finished their course yesterday and so were celebrating. The DMs did the "snorkel challenge" which

involves something like pouring vodka into a snorkel and drinking it and putting beer into a mask and drinking (inhaling??) that—they said it is part of their "initiation."

These field notes provide a glimpse into the lifestyle of the foreign volunteer tourist. These are 10 excerpts from 10 separate daily logs within the woman's first thirteen nights on the island. Ten independent references to frequenting bars across 13 nights, 6 distinct bars, and 3 separate drinking challenges. I also saw this particular woman tagged in a photo at Dr. John's house for his TNT drinking challenge; this was not mentioned in the above excerpts, but I assume it was a stop along the way during one of these nights out.

These field note excerpts highlight the divisions between locals and tourists. The bar clientele is largely made up of foreign tourists every night, save for the ninth reference. The second entry refers to Carnival week, a popular local custom, and illustrates the physical and sensory distinction between the party on the streets (attended by locals) and the party in the bars (attended by tourists and volunteers).

While volunteers often work hard to identify themselves as somewhere in between tourists and long-term expats, few locals see the distinction. In the following interview excerpt, Kathy, the then-lead scientist at one of the conservation organizations, considers the volunteer researchers her organization attracts:

> I think people expect it to be a really fun, laid-back place . . . It's interesting because with [us], it's like people know this island as diving and partying and, like, those are the two main reasons people come here.
>
> Q: *Where do you think future volunteers get that impression?*
>
> I think online. Like, there are YouTube videos. There are, like, if you do a quick search of Utila on Google images, it's all just party pictures, all the advertising . . . dive shops may be advertising that it's kind of a party island, [a] party dive shop. Word of mouth. I think a lot of people will be like, "I'm backpacking through Central America, what would you recommend?" And people are recommending Utila.
>
> Q: *And do you think reality matches it?*
>
> I think for a lot of people, yeah. Yeah, and I know . . . it's like that is a reality for us too. But then people also know that we're doing—it's such a "work hard, play hard" kind of [a] place. Like, I really like that about it

too, where it's like we can kind of still fit in that image of the island that a lot of people see, like, we are super laid back, super fun, but we also work so, so hard at our job. And like, yeah, just caring so much about the environment, so much about the island and I think we really pass that on to everyone around us.

Kathy mentions the now infamous *If You Come to Utila* YouTube video and how it helps to attract volunteer researchers who want to "work hard, play hard" but also how they are "caring so much about the environment" and "pass that on to everyone around" them. In the coming pages, I consider how this affect—this care—transforms local relations to other species. But before we get to this, let's consider how exactly this once sleepy island of Utila transformed into a site of debauchery and self-indulgence.

The Rise of Tourism Development and Conservation

Tourism has become the primary development strategy under neoliberalism in the Caribbean, helped along by policymakers throughout the region. Tourism sits center stage in Honduras' development agenda, as it does for nearly all states in the region. Honduras' push to aggressively grow its tourism economy began in the late 1980s and early 1990s through the creation of "Tourism Free Zones." These free zones qualified investors for the same fiscal benefits as Export Processing Zones, including 100 percent foreign ownership of property, municipal and federal tax exemption, and tax-free imports for industry-related material (Decree no. 98–93, 1993). The Bay Islands became one of the first Tourism Free Zones in Honduras, and, by 1997, tourism was ranked third in foreign revenue generation, just after bananas and coffee (Stonich 1999). In 1999, Congress passed another law to incentivize tourism growth, dutifully named the Law of Tourism Incentives (Decree no. 314–98). This new law granted a ten-year exoneration from income tax payments, and exemption from tariffs and taxes on imported goods and services.[15] By the turn of the twenty-first century, Utila had established itself as a backpacker's paradise with a local community that was generally happy to host them. The following selection comes from an extended interview from 2002 fieldwork with a then thirty-four-year-old descendant of one of the black islander settler families. We hear what Danny believed made Utila special, how the island changed over his lifetime, and his reflections on the state of tourism and projections for the future. He begins by talking about *freedom*.

> Utila's a special place because you know if you born here there's a lot of freedom and there's a lot of communication with the people here. We all understand each other. Actually for me it's a special place because it has beautiful surroundings like caves, reefs. It's a tranquil place—no crimes—not a lot of crimes on the island up to this day ... [Regarding change] I think you know, the reef of the island, it was extremely colorful. I mean once you got in the water, if you was local or wherever you came from, if you got in the water, you didn't want to come out. It was a different temperature and everything. It was beautiful. You know I remember as I look back I would see these tremendous coconut trees, all over the beach and the beach was completely full of sand—all over—no erosion ... The coconut trees and the big sandy beach? [That was in] the '70s. I could remember '78, I was a kid running on the beaches, playing with the coconut. You had beaches eight to ten feet wide from the water. Beautiful beaches. It sloped off, but it was always long. Right now you see what we have left ... You know those days I would go up in the coconut trees and I looked down, things looked small, you know they were huge coconut trees ... beautiful leaves and coconuts ... some of the coconut trees expanded thirty feet in width, with the leaves hanging all the way out. But now with the coconut trees we have, the most expansion we can get is ten to twelve feet, because they're sickly, you know, and the climate has changed and erosion has affected them.

Also in 2002, Naomi, a white islander, described what life was like growing up in the 1960s:

> Some days we didn't have food, but for some reason we always had basil and lemon, lemon leaves. So my mother made tea for us, and even though we didn't have food the tea took the hunger away. [*she laughs*]. Yeah, it was good. It *was* actually, I have good memories ... We depended on the boat. We had a boat that come and take our oil and we exchanged it for rice, beans, sugar, flour, all the groceries actually, and we also got some money, so that kept us going every fifteen days. I remember how I used to borrow the coconut oil to exchange with people. I think they took it to Belize and the Cayman Islands, the coconut oil.

While the coconuts declined and the beaches withdrew from the 1970s to the '90s, the reef maintained just enough diversity and the islanders just enough of their small-town, laid-back way of life, that it continued to attract foreign dive tourists looking for an off-the-beaten path and unique experience. Asked to describe the type of tourists Utila was attracting in 2002, Danny said:

> Well, right now, what we're dealing with is the backpackers. You know and one thing that I notice with the backpackers is that everybody gets an opportunity to make a little bit of money, you know. I think in times to come, we would have more sophisticated tourism that would go to five-star hotels that would be built in the future or to buy the package, but that doesn't help all the islanders because maybe 8 or 10 percent of the islanders are having these big hotels for these classy type tourists, while the backpackers is good for the whole community. You know advanced tourism only helps a resort. I think in that matter backpackers are good when it comes to all in the community.

The consolidation of tourist services that Danny worried about in 2002 did occur, and by the twenty-first century most tourists were living, eating, and partying in one of the handful of major dive centers / hotels on the island. Reflecting on the transition from a primarily agricultural and fishing economy to a dependency on tourism, Darlene, a black islander, felt the first couple years of tourism worked well for islanders:

> We turned houses into hotels or built to rent partially on the side, and it was good. I think it was going in a good direction then until the dive shops started, and once the dive shops started maybe for one year it was good with the dive shops because they would try to fill every little hotel up with their divers. Yeah. But them days they started to build accommodations for the divers, so that cut the locals out from the rentals and then they start doing restaurants at the dive shops and building bars, you know. They have this what they call "packages." Yeah, that changed it, I think. Most locals from here they didn't have the money to invest, to compete with that mentality. We weren't used to [that] here. Yeah it was a bit of a struggle on and off with the authority to try and change it . . . So yeah they [locals] continue selling their property and so on until now . . . [P]eople started switching to

the tourism. People started selling properties, you know, they offered them some money, a lot, big money.

In this next excerpt, a British man who arrived with the backpacker crowd and never left witnessed the impact that the transformation of tourism from locally owned small businesses to full-service dive hotels and resorts had on the local community and economy:

> They centralized. What I mean is they started to all have their own hotels. They all built their own bars, they all built their own cafés . . . All inclusive. Almost gated businesses. So now to the point where if—for example, my neighbor ran a bar, Coco Locos, and they didn't even go to the bar to open until 9 o'clock at night because they knew that the tourists don't leave the dive centers till then. They stay there, they drink the sunset drinks, they eat their meals. When I arrived on the island [in the 1990s] you did your diving in a dive center, you ate your food in a local restaurant, you went to one of the little bars, you stayed in a local person's hotel. Everybody in the community made money. Not anymore. Now you have ten dive businesses that make all the money, and it all goes to foreign bank accounts.

Between 1985 and 1996, the number of hotel rooms on Utila increased from 34 to 199 and several dive shops with attached lodging opened in the 1990s. By 2001, the island had eleven dive shops (more than half of which remain foreign owned) and nearly thirty hotels, ranging from modest rooms to resort lodging. The construction of a new airport in 2003 and associated highway further expanded visitations to the island. Swan's Bay, the location of the airport, was once a central breeding ground for island iguanas and nesting area for many bird species. The location also included several natural freshwater wells. This location had been recommended to remain underdeveloped due to its ecological importance and significance to the island's potable water supply (Vega, 1993, cited in Currin 2002). However, 425,000 square meters of lush secondary tropical forest were bulldozed to make way for the airport. The fragmentation (splitting of the forest) removed natural ecological corridors, resulting in the departure of several birds from the island and a decline in the iguana population. Further, erosion and sediment loss from the airport are threatening the reef ecology on this side of the island. Soon after the airport opened, twenty-four-hour electricity arrived on the island.

By 2020 there were nearly 2,500 rooms available for rent on the island, and in high tourist season 5,000–6,000 people can visit Utila in a week. During Semana Santa, many rooms fill to triple occupancy bringing upward to seventy-five hundred tourists to the island that week alone (mayor's office, personal communication, January 4, 2020). Whereas when I took my first trip to the island in 2000, transportation was almost exclusively by foot, bicycle, or four-wheelers; aircraft still landed on a small dirt airstrip next to the beach; and there was only one paved road, today motorbikes and tuc tucs travel the roads all day and night. Nagging motor beeps have replaced the scampering of blue crabs that one used to encounter on their walks along the main strip.

The old airport has been transformed into homes priced beyond the reach of most locals, even beyond the capacity of most descendants from settler families, who sold most of their beachfront property when the tourism industry launched. "They all sold and went to the states, and believe me, there ain't no way they can afford to buy their house back now," said this descendant of one of the white settler families in 2020. He had managed to remain on his family's beachfront property since their initial settlement in the early 1900s. Almost all of the other waterfront properties in the bay sold to the foreign-owned diving industry, transforming the docks and "coconut houses" (small structures where cargo boats would come and load product to sell into the global market) to scuba launchpads and training facilities.[16]

Coinciding with the growth of dive tourism and housing development came protected area management, both of which had a direct impact on local livelihoods. As Fannie, a white islander, shared in 2002:

> Look, the country didn't care about Utila for years, but now tourism is coming. They all of a sudden are being recognized. Before it was all Roatán, Roatán, Roatán—never Utila . . . Now they're telling them to stop being fishermen. All these Americans saying "stop doing this" and "don't do that," but the first thing they want to eat when they sit down is a lobster or a big fish. Look, this is always been a poor fishing village. The government should give the islanders a break—put islanders to work. The hippies are gettin' all the jobs. They get the top dollar and they get trained in the dive shops.

Fannie is referencing the establishment of BICA, the first conservation NGO, mentioned in the book's introduction (and discussed in more detail in chapter 3). As previously noted, the origin of BICA aligns with Honduras' push to aggressively grow its tourism economy in the early 1990s through the creation of Tourism Free Zones and associated accords to encourage tourism development in the Bay Islands.[17] New legislation and management strategies were introduced to conserve the area's biodiversity, especially in areas with tourism potential. Conservation and tourism development became forevermore married on the island, and from this point on, tourism, conservation, and volunteers have assembled together to reshape multispecies encounters.

Of Mangroves and Money: Development, Dispossession, and Deforestation

> There's been a lot of construction. A lot, a lot. Before . . . If you wanted crabs we would go to Pumpkin Hill because it was incredible the quantity of crabs there. In September, it was incredible, you could see thousands of crabs and grab a bunch, but now there aren't any. With even this you can see how much the island has changed, just by looking at the crabs. You can go swimming and see nothing . . . I think about it you know and, and when I was growing up we had crabs in the nature. Could sit down and watch the crabs going in to wash their spawn in the sea. Huge amounts of crabs. I don't see that anymore. Of course, the island had changed 'cause on the seaside is where they would go to wash their eggs out, but there are homes there now. So the crabs—they can't get to the ocean like that easy.
>
> —Mainland Honduran who emigrated to Utila in the early 2000s when she was in her late teens, reflecting on decline of crab in 2016

As tourism and conservation took off, so did housing development and an increasing number of foreigners (mostly Americans) seeking Caribbean getaways. Three Realtors emerged on the island (none of them Utilian) and boosted the real estate market, even bringing global attention through the aforementioned appearances on HGTV. Development was helped along by a series of changes to land legislation. While conservation codes were put in place to protect endemic species and critical ecosystems, there were always workarounds for those looking to work around.

Mangroves are continuously being cleared and infilled for housing expansion, which has a direct impact on local food sources. Conservation professionals are well aware that their biggest struggle is not islanders and their tastes for turtle, crab, or iguana but the threats to habitat survival created by construction to accommodate the growing tourism industry and international house hunters. In this section, I share three examples of dispossession, showing how some island inhabitants—both human and other species— have lost the ability to dwell within their land and mangroves, while other residents move in. These are just three examples of countless cases that islanders have shared with me over the years.

Tradewinds

In 2001 I stayed in a small hotel owned and managed by two Utilians who also kept a home in New Orleans.[18] The woman managing the property was selling her mangrove lots that met up with the old airport road for $10,000, and she tried to convince my husband and me to invest in the island's imminent development. She saw the writing on the wall, and the changes that were about to occur with the plans to move the airport off the beach and bring twenty-four-hour electricity to the island through a private company. The changes happened fast, and the scale of real estate inflation is made clear comparing the $10,000 price tag that Fannie put on the land with what it is valued at now in the same area, currently known as Tradewinds (figure 11).

In 2017, a small, one-room studio in the area Fannie tried to sell us was listed for sale at $231,000. It was one of several foreign-owned properties that had been featured on the HGTV's *Caribbean Life*, season 4, episode 14, "Lawyer Trades City Life for Relaxing Beaches in Utila" in 2015. The description on the realty site in 2017 stated:

> This well-constructed home is located just past Bando Beach on the Point in Utila in an area referred to as the "Old Airport," as the first small aircraft landing strip for Utila used to be located nearby. Featured on a recent HGTV "Caribbean Life" episode, this property is about a 15-minute walk into the centre of town and an easy and quick bike ride to all the amenities. Enjoy views of the Caribbean Sea as it washes against the eastern shore of Utila and the dive boats that pass by. The sea is only 100 yards away [and] the vibrant

Figure 11 Tradewinds homes with mangroves across the road. Prior to construction, mangroves reached the shoreline. Photo by author.

coral reef can be easily accessed for snorkeling as can multiple world class scuba dive sites—many marked by dive buoys—right from the shore.

The only access for snorkeling or diving from this house is to walk on dead and dying coral and seagrass, something even a novice would know is problematic. This particular house seems to remain eternally for sale. At the time of this writing, its price had been reduced to $165,000. Still too much for most locals.

The emergence of Tradewinds was controversial at the time and allegedly corrupt, but many younger Utilians do not know the history of conflict nor what the area looked like twenty years prior. The environmental license for construction in the area was approved in 2000 and included regulations with respect to mangrove clearance. Once development began, the properties were audited on several occasions, in response to community complaints that alleged the leveling out of the coastline and clear-cutting of mangroves. For instance, reflecting on the largest environmental problems facing the island in general, and the swamper in particular, a Utilian conservationist shared that the housing developer claimed to not know about the endemic species or the impact of his development on the mangrove and beach ecosystems.

> When they [expatriates looking for vacation or retirement homes] buy and sell land, they cut mangroves. This affects all the mangroves [and] the fisheries, [and] the nurseries . . . The mangroves give us support for the island when the hurricanes come . . . [But] because the people want the typical house on the beach, they cut the mangroves. They want a "paradise" house and they don't know what they are really doing . . . But some of them know the consequences of cutting the mangroves: this guy from Tradewinds [a housing development], he was like [motioning with shrugging shoulders, hands up] "I don't know. There was no mangroves here. Hurricane Mitch cut them all, took them off." And you're like, "No. You see that picture from past and present? You see the difference." And he did it. But only he's like "I don't know anything. I didn't know we had an endemic swamper. I don't know about the mangroves."

According to the above quote, the Tradewinds developer denied making any adjustments, stating that recent hurricanes are responsible for the transformation of the coastline. However, part of the audit process includes taking before-and-after aerial photos which are now on record with the municipality. While photographic evidence supports allegations of infilling, the auditing process has no real teeth. The developer is fined and "educated" (again) regarding construction guidelines. Often fines are so minimal that they do not dissuade developers from misconduct.

This Old Head pointed to local politicians and land surveyors as the cause of dispossession:

> A gentleman come from the United States, some part of the States . . . Perhaps you've heard of him? . . . The people criticize this man. As a bad man, he's no good. That's not true. People don't speak the truth. He came and he bought a piece of land . . . Who is to blame? For this land problem, is mayor and the surveyor. They [are] the two to blame. Because the surveyor has instruments to tell him that this land has already been surveyed . . . And if he's come to this piece of land, that's in his logging. Whenever he goes to

measure a piece of land, join in this, he know exactly where to commence and where not to commence. But he goes and takes three hectares of [Fannie's] land . . . And he joins it on [the developer's] piece. And he makes the cropped pieces, and takes it in to city hall and the mayor signs it. But half of the land owned to [Fannie]. And this is facts . . . And the surveyor, for the love of money, knows that he's into this lady's land. But he don't care because all he want is the 10,000 or 20,000 lempiras that they [are] going to pay him for the measurements. Which is only probably a couple hours work.

This man recalled that he believed Fannie was contesting the land loss in court, but in his opinion, it's not the foreigner to blame for land dispossession. Rather, it is islanders, one's neighbors, and maybe even family. He continues:

It's like this: You the stranger here. This is mine, if you come and you buy a piece of the land joining me, and the surveyor comes and measures half my land. And put it on your documents. They can't call you a thief. Because you didn't take nothing. It's them that took it and sold it to you. They're the ones that stole—they the thieves. They're the ones that sold it to you, is the thieves. Not you. Because you don't know who the land is for or where's it at.

How did foreigners get license to develop in the first place? Take the next case of Bando Beach.

Bando Beach

Bando Beach is located on the old airport strip, at the very end of the island in an area locally referred to as "the Point." It is a private beach restaurant associated with a nearby hotel, located close to Tradewinds. Tourists must pay a fee to access the beach; locals do not. Back when the area served as Utila's airplane landing strip, the very small beach at the edge of the strip was a public swimming area and snorkeling access point. The Bando Beach example illustrates both the minimal impact auditing a development can have in addressing concerns about the environmental impacts of a development, as well as the privatization of once communal resources. The Bando Beach case also illustrates two workarounds to Honduran law: one, the fact that Honduran land cannot be sold to a foreign person, and two, that beaches cannot be privatized.

> What happened with Bando Beach ... [is] this doctor got married with a local woman, so they got this piece of land. And over there used to be the airport and once they sold that part of the airport, they took part of that land. And it used to be all mangroves on the edge of the coast. So what happens is they created this Bando Beach idea and everything and ... they couldn't take care of it. They said people were destroying it and this and that, so they decided to fence it. And even though we tried to [audit]—because according to the one agreement you cannot fence—but it's inside an urban area, so [it turns out that] we have no say on that.

The "agreement" that this conservationist (the same person as was describing the Tradewinds development) refers to is Article 107 of the Honduran Constitution, which prohibits non-Hondurans from owning coastal land. However, since the doctor married locally, together they could purchase land under the woman's name. The speaker then references Decree 90/90, the reform that enabled foreigners to purchase "urban land," and the subsequent ruling that classified all land suitable for tourism development as "urban."

While under Article 107 of the Honduran Constitution, foreign-born individuals cannot own land within forty kilometers of the coastline or borders, in 1990 exceptions to this rule were put in place through the passage of Decree Law 90/90. This law stated that urban land could be owned if the property were deemed to have social, economic, or public development interest by the secretary of tourism. In 1992, the Rafael Callejas administration passed legislation that declared all land suitable for tourism development to be classified urban land, hence opening all of the Bay Islands to foreign purchase.

The municipality issued a complaint to mainland authorities regarding the development of Bando Beach, and officials came to the island to investigate. Because the fence was on designated urban land and the owners stated they were charging a fee solely for maintenance of the area and not the beach, there was nothing further the municipality could do. Under Honduran law, if someone enters the beach area from the water (swims or boats over), the owners should not be able to run you off the beach; they only have authority to remove someone if they use something of the owners' (e.g., a beach chair, water spigot). However, mainland tourists and foreigners are charged a $3 maintenance fee to enter. Moreover, in 2011 people suspected that the owners were bringing sand to fill in the beach and, thus, heightening erosion. Having frequented this small beach regularly between 2000 and 2002, and then seeing it again on a regular basis from 2011 on, I concur that it had more than tripled in size by 2011. Every visit there since 2011 seems to be met with more sand.

The Camponado

Upscale housing developments and private beaches are not the only reason mangroves have been cleared. They were also cleared to provide space for the rising mainland population in the Camponado. When mainlanders began to migrate to Utila in the late 1990s, they were constructing small shanties in the inner lagoon, but also in other parts of the islands, including areas that original families owned and lived in. In 1998, the municipality decided to allocate the wetlands Camponado land in what someone from the municipality described to me as an "organized" fashion and to sell it in a "symbolic way, at a very low cost" (reportedly 8,000 lempiras, or US$615). Many took advantage of the low land cost and purchased plots, including (nonmainlander) Utilians; today plots continue to be filled in and homes constructed, and plots are even resold. The following excerpt from a joint interview with two conservationists, whom I will call Nanci and Tom, reveals some of the extreme environmental and health problems associated with this construction (figure 12).

> **Tom:** The problem with Camponado is that they never put the water pipes [in]. Like all the management of the black waters and gray waters, and that's a bomb for Utila. You sometimes can walk on the main road in town and you can feel the smell, and it comes from Camponado.
> **Nanci:** And if you ever look at the children; a lot of them have very bad skin problems. Sores, cuts, infected cuts, and that sort of thing.
> **Tom:** And the problem is they throw their own trash and with that same trash, they fill the land.
> **Nanci:** And that includes car batteries—
> **Tom:** Bicycles.
> **Nanci:** Fridges, the freon.

Figure 12 Camponado lot under construction for new home. Photo by Josely Turcios

A third local environmentalist interviewed at a separate time shared:

Rose: The issue with that is it's mangrove; it's water. So people have to be filling in or building their house on poles... And the problem is that with the trash—you see it on the side—we have tried to clean, but it's their filling, so you can't take it out. And if you try, they get mad at you and they run you [out]. And all the plastic... They just put it in and it starts to compact. So they use it as filling.

Returning to the joint interview, when I asked them if anyone was studying this environmental injustice, I was told:

Tom: It's a bit of a hot potato... The problem is people have been sold land. Whether it was incorrect or correct, they have paid for their land, so they have all their legal papers for the land. It was condoned and allowed to happen by the municipality... and you could do a study, but at the same time, what are you going to do with the study?... Definitely it would expose the situation but at the same time, what really is going to happen? At the end of the day, probably the only solution is—the only realistic solution—is to take everyone out of there. And so then you're going to displace the poorest people that are living on the island.

Yet, global economic pressures, political instability, and social unrest caused by the 2009 coup d'état continue to push rising numbers of migrants from the mainland to the island in search of labor.

Dispossession, Gender, and the "Gringo Mentality"

International travel and housing markets have transformed any available land into something no longer attainable for most islanders (outside of the few original families), confining increasing numbers of Hondurans and early Utilian descendants to the Camponado and similar fringe developments. And, local conservationists are concerned because the small penalties associated with breaking environmental legislation are not enough to deter further development in fragile zones. According to Tom,

> The American goes, "I can cut it down and then what? They're gonna fine [me], what, a thousand lempiras (US$53)." And that unfortunately has been the gringo mentality. They've been told here by other gringos for the last ten years, "Just do it and then worry about it after." Because it's cheaper to be fined and just go through the process, than it is to apply at the beginning and just get your permission. So even the system itself encourages people to behave like that... Because the whole gringo mentality here is "do it and pay the fine after because it's cheaper."

What Tom calls "a gringo mentality" is part of the moral philosophy of neoliberalism. As the state withdraws, individuals are increasingly asked to take care of their own needs, and islanders begin to "sell pieces of paradise" to foreigners, some of whom could care less about mangrove preservation.

The accounts that some islanders have shared suggest dispossession was gendered, with politicians enabling land sales when men were working off the island, on ships or in the cargo industry on the mainland. Rumor from one of the Old Heads was that the Camponado belonged to one of the early settler families. Speaking of the woman who owned the land, this man shared, "She had all girls, and neither boy. Some of them went on the mainland, some went to the states. But there was only [names two women] here. They stayed for the farm. [Name removed] was an old woman then. When [certain politicians were in office, they] took Camponado. Camponado belonged to [name of a woman removed] then. All of Camponado. But he took it from them, because there was no man stand(ing) to defend them. Only women." The same Old Head shared how the rest of the land along the bay and east end was sold, all of it to him, by exploiting women on the island. "All that he sold. He left those old women's walking 'bout Utila streets barefoot. And in rags. And with no food. While other people was, just . . . it was just sold for millions of lempiras. Millions! I tell you, that's a crying shame! These the kind of people we don't need! Don't do these kinds of things."

This man was not alone in documenting the gendered nature of land loss. While he was one of the only to call it out as specifically gendered, women's names featured commonly in stories of land loss.[19] Speaking of land loss in particular, Mr. Austin, the man who told me the earlier accounts of attempted rape, said: "This island, things [are] hidden from you. Because they don't want you to know the truth." Just as William Jackson documents in the *Sea Shall Hide Them*, Utila is a place full of secrets. It is a paradise that enables some actors to act with little to no repercussion.

Remember W. E. B. Dubois's (1915, 12) famous quote about white man's privilege—"to go to any land where advantage beckons and behave as he pleases"—and its application to the tourism industry. Most certainly, this applies to Utila and is captured in this video where tourists are transforming the island into their playground, because as the lyrics say, "you can do whatever you please." The effect of this privilege is increasing dispossession, both rhetorical and material, and the inability to live freely in the ways one's ancestors did.

There are striking similarities between the representational patterns of Caribbean plantation landscapes and today's tourism advertising of Caribbean "paradise" (Feldman 2011). In the former, the slave labor upon which plantations rested is omitted, with the plantation an imaginative place in which Europeans can effortlessly attain wealth, reflecting a one-world-world framework based on capital accumulation through dispossession (Escobar

2017; Feldman 2011, 47; Strachan 2002, 41). The real labor upon which that wealth is produced is obscured, hidden from view through the presentation of a seductive paradise lost, where slaves—if even present at all—appear as content, quaint accents to benevolent masters. Likewise, with tourism advertisement, the labor behind paradise disappears. Feldman (2011, 47) describes some of these omissions from advertising: "single mothers working long hours at resorts for dismal pay, farmers selling land under pressure from developers, government-sponsored 'tourism awareness' campaigns." Instead, tourism advertisements present the Caribbean as "a timeless, workless, pleasurable utopia, inviting European and U.S. tourists to take a break from work, slow down their pace of life, and temporarily join the endless leisure."

This pleasurable utopia is precisely what Utila's young tourists and volunteers are seeking, yet it obscures the realities of Utilians who live in this space and who have seen their island transformed by increasing numbers of tourists, lifestyle migrants, and migrant laborers. The YouTube joke that "you're never gonna leave" is far from true. Sure, many tourists will extend their stay, and while many of these tourists won't do substantial damage to the ecosystem as individuals, the cumulative effects are disastrous. But for the lifestyle migrants who arrive and clear the mangroves to build McMansions along the coastline, they soon discover that life is hard on Utila. True, it is a beautiful Caribbean island. But it is still Honduras. And life in Honduras is not the same as life in the United States, no matter how closely you mirror your physical dwelling to "American standards." In January 2020, Marta, the Garifuna islander quoted earlier, shared her frustrations with a snowbird in Tradewinds who was complaining about how thick the mosquitos were in October (an unusual time for mosquitos to proliferate but brought on by warmer climates associated with climate change). The snowbird said to Marta, "I don't have to put up with this. I have a U.S. passport. I'm going back to the states, and I'll come back here once the mosquitos clear up." The woman up and left, leaving Marta and the rest of islanders on the island among the clouds of mosquitos without the freedom to up and leave "whenever they please."

Honduras, for many tourists and lifestyle migrants, appears to be an eternal Caribbean paradise. For temporary visitors and new homebuyers, its attraction seems as if it can never end. But the "for sale" signs throughout the island suggest otherwise. The vast majority of newly constructed houses seem to go up for sale within just a few years of being built, and as you walk through town you will see more houses and businesses with for sale signs than not.

To summarize, this chapter detailed one of the two main currents of the book: specifically, here we read how neoliberal capitalist pursuits to grow a tourism economy that favors ideas of "freedom" and "discovery" have dispossessed Utilians twice over. From the decimation of the Pech to the contemporary erasure of descendants of settler families, a new way of worlding entered and significantly transformed the landscape. With the colonial erasure of the Pech, establishment of a plantation economy, and settler ideologies of puritan order, *some* ideological threads of the hegemonic one-world world shaped the ways humans and other lifeworlds assembled in and around Utila. However, the one-world world was not *the* only or perhaps not even the dominant form of worlding. As we saw in chapter 1 and will see in the next chapter, early Utilians engaged—and continue to engage—in reciprocal relations with the nonhuman world, assembling together in a relational form of worlding practices. Now—today—the pursuit of pleasure and discovery make for additional newly emerging landscape assemblages, with shifting boundaries, as both humans and other species claim, reclaim, and reconstruct boundaries and territories (Latour 2007; Ogden 2011; Ong and Collier 2005; Tsing 2015). The next chapter picks up some of the elisions in the conservation and development narrative of Utila, moving to the second main thread of the book: the exploration of conservation voluntourism as an industry that is significantly transforming material and affective relations between and among species, human and more-than-human, through a tourist-centric worlding framework.

Chapter 3
Conservation, Volunteering, and the Spectacle of Affective Labor

> We didn't know too much about the Whale Shark. We knew about the Old Tom. We called him here the Old Tom. So the Whale Shark came up since the tourists been coming around.
>
> The reason why the fishermen are so interested in protecting Old Tom is because the Old Tom is the life of the fisherman. [If] there's no whale shark, there's no bonito. When the whale shark go, he take all the bonito with him.
>
> —Quotes from two different Cayan fishermen in the documentary film *Big Fish Utila*

The mariner and agricultural livelihoods that became a touchstone for the formation of Utilian subjectivities would be forever transformed with the first backpacker arrivals and launch of dive tourism. Before the tourists arrived, Old Tom had no other name. This ginormous whale friend stewarded the waters and brought with him good fishing. Whereas in some parts of the world people harvest their meat and oil, Utilians never engaged in this market. For Old Tom brought with him the bonito. When Old Tom is present, that means bonito tuna are present feeding on plankton. Large boils in the water are a good indication that whale sharks are near. Today, tourists come to Utila to swim with the whale sharks, a new source of revenue for local mariners but one that comes with its own problems, from ecological degradation associated with increased population pressure and infrastructure development, to a widening wealth gap and shifts in the owners of capital from local to foreign, to sociocultural changes brought by the intermingling of people from wide-ranging backgrounds. Along with the increasing pressure on the local ecology that accompanies the tourist search for freedom and mainlander immigrant search for livelihood is its counterpart: heightened environmental research on ecological change.

Anthropocene Tourism, Conservation Voluntourism, and Nature on the Move

Amelia Moore (2019, 6) suggests that the "Anthropocene idea" has inspired new entanglements of industry that we have historically thought about as distinct, and that these new entanglements remake the world to benefit certain international industries at the expense of local peoples. The term "Anthropocene" was coined by chemist Paul Cruzen and biologist Eugene Stoermer in 2000 to refer to a new epoch when human actions and activities began to influence and change the entirety of the planet. The idea of the Anthropocene thus calls to action a suite of experts to understand and respond to the environmental changes brought on by human activity and the associated threats to our and other species caused by recent biological, chemical, and geologic transformations. Moore (2019) refers to the constellation of research studies on the Anthropocene as "Global Change Science or GCS."[1] She points toward island places (e.g., the Bahamas, her field site) as laboratory experiments for global change science studies. Utila is another laboratory.

Island tourism has long been a part of national economic development plans in the Global South. In the current age, tourism in these locales is being refashioned and rebranded from simply a "sun, sand, and sea" vacation to one that combines relaxation with adventure and environmental stewardship. Moore (2019, 21) explores this transformation in the Bahamas, observing that while tourism has featured prominently in Bahamian historical and national development narratives since the mid-1900s, today "Anthropocene narratives are starting to intersect with tourism narratives." She argues that since GCS has identified places such as the Bahamas as "vulnerable" and in need of protection, the tourism package is therefore now sold through a narrative of environmental protection. Moore calls this tourism approach "Anthropocene Tourism." Here and elsewhere I refer to this tourism strategy as "conservation voluntourism" (Brondo 2013, 2015, 2019).

A paradox emerges as tourism officials anchor the "tourism product" in the vulnerability and nature of islands themselves and as enhanced environmental protection gets linked to increases in tourism and tourism revenue. This is the central paradox of the twenty-first-century green economy: it is an economy that pushes for the continued expansion of global tourism as a means to mediate the environmental degradation and social inequalities that the industry itself creates. Studies of this paradox and the industries that support it are still underdeveloped. As well, both studies of "science tourism" and research grounded in a political ecology of tourism remain few and far between, especially by anthropologists (Moore 2019, 63; Mostafanezhad et al. 2015).[2]

Analyzing emergent assemblages is part and parcel of today's anthropological scholarship. Whereas early anthropology approached culture and the "field" as bounded entities, and anthropologists focused their attention on describing a discrete, homogenous culture that they encountered within a clearly bounded space, today we acknowledge the fact that culture was never neat, bounded, homogenous, and locatable in an exacting way. In the past, tourists did not "fit" within anthropological studies; they were intruders on the neatly bounded categories that anthropologists were documenting. Now tourists can be seen and reckoned with, understood as part of an assemblage of beings that create cultural meaning. This book is one small attempt to analyze the ways in which tourists, global change scientists, islanders, conservation voluntourists, and nonhuman species assemble together and produce new subjectivities.

Conservation voluntourism (or "science tourism" for Moore) involves the creation of new subjectivities. The industry is based on an imagination of a particular type of student and earth. Moore (2019, 63) writes that within this industry, "The planet and student subjectivity are malleable, improvable, and receptive to positive change based on knowledge." For the tourists who participate in conservation science, they are presented with a place (e.g., the Bay Islands or the Bahamas) as comprising local people, marine and terrestrial species and ecosystems, and natural and anthropogenic change that is uneven but can be managed through science (Moore 2019, 63). The language of science, then, is enlisted as truth on which to base "good" policy. Conservation volunteers become enrolled in the making of scientific reality, collecting data that are then transformed into policy with real, material effects for local populations. Clearly this activity has implications for not only local livelihoods but also the individual subjectivities of "hosts" and "guests," as well as for species survival and proliferation.

In places such as Utila and the Bahamas, local people become research specimens within the landscape of GCS fieldwork and conservation voluntourism. People are surveyed about their local knowledge of species and schoolchildren become targets of environmental education programming, transforming local people into objects within the tourism product and conservation experience (Moore 2019, 22).[3] This experimentation with people carries forward the colonial practices and assumptions of authority of specific places, people, and actions, an extension of the one-world world. GCS that involves conservation volunteers detracts from the labor and value produced through grounded local interactions with natural resources. In this chapter, I argue that promoting and bolstering the conservation voluntourism industry may end up advancing the creation of new neoliberal citizens while further concealing the micropolitics of commodified nature.

Capitalism, Spectacle, and Multispecies Ethnography

Conservation voluntourism, like ecotourism, payments for ecological services (PES), carbon trade, species banking, and other market approaches, has emerged out of the material transformation of nature within the context of capitalism. That is, under capitalism, nature has come to be seen as "a bountiful pool of resources that exist either in the form of material resources, or more recently, in the form of services that are meant to satisfy human needs" (Neves 2010, 726). Conservation voluntourism, nature tourism, and ecotourism are service industries that create opportunities to satisfy human emotional and psychological needs.

The process described in the book's introduction, through which nature becomes valued as a commodity, appearing in the market as if by magic, detached from the social and material labor that produced the service, is described by Marx's concept of fetishization.[4] Under multispecies conservation voluntourism, when flagship, endangered, and vulnerable species are reduced to commodities to serve nonmaterial human emotional needs or experiences desired by (mostly) western millennials to be competitive for employment in the global economy, the social and ecological costs of voluntourism are obscured from view (they are fetishized). For example, when tracking and monitoring whale shark movements became valued as a tourist attraction that contributes to the preservation of the species that enables voluntourists to encounter whale sharks up close and personal, the sharks' value to the island fishing community as locators for tuna diminishes, and the ecological costs of increased travel and development pressure on Utila are hidden from view. Katja Neves found similar fetishization in her work on cetourism.

In her work, Neves (2010) demonstrates that whale watching, like many other forms of multispecies ecotourism or multispecies conservation voluntourism, presents nature as a service provider to be consumed in situ by tourists. The promotion of whale watching by international environmental NGOs and tourist companies alike obscures the contradictions of nature and capitalism. Neves (2010) argues, for example, that by equating whale watching with ecologically sound cetacean conservation, tourists and even conservation NGOs themselves reduce their ability to recognize instances where whale watching actually ends up causing more harm, for instance, through underwater noise pollution, the stress of excessive numbers of boats around, and humans swimming with cetaceans.[5] Whale watching, like the forms of multispecies voluntourism discussed in this book, presents opportunities for humans to connect with nature, to experience and "become with" other creatures, and, presumably, through this capitalist endeavor of consuming such experiences, they will solve the problems created by capitalist extraction in the first place.

Thus far we have seen how Marx's work on commodity fetishism can be applied to conservation voluntourism to understand how the industry creates value in the trade of experiences in or with "nature" while detracting from the labor and value produced through grounded local interactions with nature. This new value is what underscores contemporary multispecies voluntourism and is central to Honduras' "affect economy." I introduced the term "affect economy" in the book's introduction to describe an emerging economic system based on the production and exchange of emotional services attached to "becoming with" another species (Haraway 2008), and the subsequent cultural capital that emerges for the voluntourists.

What exactly might it mean to become with within a multispecies conservation voluntourism context? The multispecies literature conceptualizes relationships between humans and nonhuman animals relationally, theorizing the body as if in a dynamic "dance of encounters" (Haraway 2008, 4). Beings, therefore, become with, rather than simply statically *be* in the world. As noted in the book's introduction, this form of postdualist inquiry pushes beyond a narrow and anthropocentric understanding of the human condition, extending the ongoing deconstruction of the Cartesian human/nature divide with a shift toward thinking of materialities not as passive "things" but as active entanglements of many entities, organic and inorganic, that are constantly in a state of emerging. While multispecies conservation voluntourism may present possibilities for emergent ecologies, in the book's conclusion, I speculate that most multispecies conservation voluntourists actually struggle to become with, and this is because they have been conditioned by what Guy Debord ([1967] 1995) terms "spectacle."

Spectacle refers to the rising commercialism supported by the widespread dissemination and mediation of commercial images that lack content and conceal inequities and conflicts, through which masses of people have become conditioned. Through the presentation of such spectacular media productions, individual subjectivity disappears and is replaced by a singular market consciousness (Debord [1967] 1995). In the realm of twenty-first-century conservation, nature as spectacle is visually articulated and circulated through media presentations of celebrities, corporate leaders, and high-profile conservationists sharing a message that capitalism is the key to future ecological sustainability.[6] This worldview goes largely unquestioned because its cause is also the solution: society as spectacle, like nature as spectacle, is produced and reproduced by capitalist relations, as well as offered as a solution to the negative impacts associated with advanced capitalism and toward financial and ecological sustainability (Igoe 2016; Igoe, Neves, and Brockington 2010).

A large body of work in anthropology reveals how the humanitarianism industry, and associated volunteerism, rests on images of a depoliticized and dehistoricized suffering subject or category of person (e.g., hurricane victim, orphan, AIDs patient, etc.) that are meant to generate compassionate responses.[7] Along the same vein, the multispecies conservation voluntourism industry presents other forms of suffering subjects—that of the threatened whale shark, iguana, or sea turtle, for instance. Just as in humanitarian travel in which volunteers are compelled to travel to the Global South to participate in and then publicly perform their care work through social media posts and blogs, volunteers in conservation are called to address the immediate danger to biodiversity through images of beautiful, threatened species, which legitimize volunteer intervention. Both traffic in images of the suffering of innocent others and the photos of orphaned children in need raise similar moral imperatives to intervene as do photos of the non-human animal victims of human predatory behaviors.

In what follows, I explore how the value produced through the emerging multispecies voluntourism industry alters the material and affective economies and worldings that existed on the island prior to the growth of dive, nature, and conservation tourism. The multispecies becomings in the conservation voluntourism industry consist of new material and affective assemblages as conservation voluntourists purchase these experiences and exchange this accumulated cultural capital for a new social status. I build up to this argument through a look at some of the programming of local conservation organizations.

From Local Management to Tegucigalpan Leadership at the Bay Islands Conservation Association

The Bay Islands Conservation Association is the longest-standing organization in the Bay Islands, founded in Roatán in 1990, with chapters in Utila and Guanaja opening soon after. BICA was formed to "initiate and coordinate efforts in protecting the islands' fragile natural resources" (wearebicautila.org). To that end, BICA Utila has coordinated several projects in its thirty-year history, including sea turtle monitoring, beach cleanups, reef health checks (coral bleach monitoring, reef fish), water quality management, installation of mooring buoys, environmental education programming, ecoleader training, recycling initiatives, lionfish awareness and lionfish derbies, protected area patrols, and the development of codes for diving and whale shark encounters (together with dive shops and other conservation organizations). Many locals despise BICA as it represents a threat to freedom to move about and assemble with the natural world in ways one's ancestors have always lived (as described in chapter 2). While this sentiment has existed since the inception of the organization, it was not nearly as widespread as it is today, possibly linked to a change in leadership but especially linked to neoliberal conservation.

BICA's first director was no other than Shelby McNab (chapter 1). McNab holds a warm place in the hearts of most islanders,[8] and back in the days that he ran the organization, islanders felt a bit more attached to the organization and its programs. Mr. Shelby once shared with me that he was inspired by the island's history as the twenty-seven-year refuge for Robinson Crusoe to conserve as much as possible of the island. BICA's first programs were not so well received. Their first big initiative was to curb islanders from removing sand and gravel from the beaches. In the early 1990s, concrete mix for construction was not being brought over from the mainland, and as tourism was beginning to transform the island and leading to new construction, islanders would haul sand and gravel in sacks from the shores to construction sites. BICA wanted to stop this removal, and islanders were not pleased. So, in time, the organization worked toward a solution that islanders might buy into, literally. The question was: How do you get materials over to the island in bulk and at a price that people would be willing to pay to avoid the manual labor of removing sand by hand in sacks? The solution was to support the importation of concrete mix from the mainland, which seemed to work to dissuade sand removal. This compromise attended to islanders' dual interests in having access to small beaches for weekend leisure and affordable building material, while also serving BICA's objective to slow the pace of erosion caused by sand mining, and at the same time promoting nature tourism.[9]

Another early project that rankled islanders was the installation of buoys for dive boats to anchor without impacting the reef. With the emerging intertwinement of conservation and development, in-water experiences began to privilege the tourist's interest in diving over the local fishing economy, and BICA worked to accommodate the growing dive tourism in a way that would be the least ecologically hazardous, by installing permanent sites to anchor at the primary dive sites. The initial buoys were installed with a Styrofoam ball tied to a very long rope that anchored down. The dive boats could then easily locate the Styrofoam ball, lift it into their boat, and tie up to anchor. One white male islander who was part of BICA at the time shared the result: "But the problem was that the islanders' propellers would get caught in the yellow rope floating on the water so it was not a good idea for the islanders. So, every buoy that BICA put up, they cut it!" Early BICA members then worked to create a variation to the buoy, this time using cylindrical PVC pipe that anchored to coral pads and stuck out of the water by two feet, with a U-bolt for boats to easily tie onto. The same early BICA member shared, smiling in delight as he recounted,

> And once they came up with that design [*clapping hands happily*]—well that was the perfect design. Fisherman then liked it. They didn't like us with the other buoys. They cut every one. They were just confrontation all the time because it was hurting them, okay? Meanwhile, once we put in the new type of buoy they, their boats could—you know—'cause they have the long dories. The dory would go slide right down the side. Even if they hit it, it would just go to the side and it was not a problem, and they loved it because in case of wind or storm they could easily tie on and have an anchor. So it became

very acceptable. So all of those little things, even when you're working with the environment you have to figure out what works for the community also, because if you're just fighting people, you're not really accomplishing a lot.

Around the time that the buoys were being installed, BICA introduced the turtle project (figure 13) described in chapter 1. Part of the intention of the turtle project was to raise awareness about garbage disposal. For generations, Utilians would throw their trash in the sea, as they lived minimally and the majority of their trash was biodegradable. With the move to plastics and hazardous packing materials, trash disposal was becoming a problem. Describing the ways in which the turtle project helped to translate other environmental lessons to islanders who were once overtly resistant to BICA's efforts, this early BICA member reminisces:

Because of the Turtle Project we were making people aware, aware that you can't be throwing everything in the water. Contaminating the reefs and stuff like this. It's an evolving thing from being a very sleepy little community into a [*chuckles*] now getting busy with people coming and you know, visiting tourists coming to dive and stuff like that. So, [the man who launched this project], one day came out with ninety baby Hawksbill turtles that he found out of a nest. And [he] brought 'em into a big dishpan, [a] big blue dishpan I believe it was. And, well the kids learned about it, and man it was like a rallying movement! All the kids in town wanted a baby turtle! And he come out with a brochure or a little page with all the information that you needed—how to feed it, change the water every day. And here are these big macho Utilians that were giving us a problem [*chuckles*], okay? Cause we were dealing with some problem fishermen, stuff like that, and some of them were kinda rude to us. And the parents then, well, during their three months that they had to keep them, they would have to go down to the water with a bucket and gather up a bucket of saltwater to change the water in their tanks or in their basins or whatever. And we went around inspecting and helping and you know, and teaching them how to raise them. And the kids would go, have to go catch little minnows and frys and different sardines and all kinda little clams on the beach and rocks and stuff to feed their baby turtles. So we got the islanders involved in conservation in a very simple way by just working with the kids, with their children, so we got the mothers and fathers and oh it turned out to be a beautiful project!

Figure 13 Framed photo of the 1990s turtle project, positioned in a central location in an islander's home in the Camponado. Photo by author.

They continue:

> Well anyhow, after we raised them for another, what, eight months or so, they became like a year, yearlings. We weighed them. We have photographs of all the information. We released them in '93. We took the big turtles to Blackish Point.[10] We had a big party! All the tourists were invited, all the parents, and the kids. And we put them on the beach, okay? And they had to crawl like if they were little babies. I mean the turtles. And crawl off into the ocean. So it was such a beautiful introduction to islanders of something you can do to help the island. And it was educational. Here was an educational project working with the children, and that's how you can change family or an adult. It's hard to deal with a person already set in his ways if you don't have some kind of educational program to change. And so the kids would say later, "Mommy, don't be eating any turtles." Or "Daddy, don't be eating any turtles, that may be mine!" So it had a psychological effect. And now those same kids [now adults] are our best supporters!

Mr. Shelby was the BICA director at the time of this turtle project, a project the organization did twice during his leadership, until it was told not to any longer since the retention of sea turtles is prohibited under the

Inter-American Convention for the Protection and Conservation of Sea Turtles (described in chapter 1, note 2). Many islanders I have met over the years still bring up this project with me, sharing fond memories of their time with their own baby turtle. Several family homes have framed photos of their participation in the program, showcased proudly in central locations in living rooms (see figure 14). Several photos date back to the initial program from 1992. Many islanders feel that BICA hasn't done anything of real significance since the turtle program, and today few children are able to experience the release of baby turtles. This activity is now largely attended to by volunteers, locals who live near the nesting sites, and BICA's Reef Leaders.[11] The Reef Leaders accompany the BICA staff member in charge of the turtle-monitoring program. Through my own work, I have participated in turtle exhumations and releases on several occasions (figure 14).

Over time BICA's Board of Directors (in the words of a local) "pushed Mr. Shelby to the side," replacing his leadership with biologists and other scientists from the mainland.[12] Islanders who had once been more active in BICA found that "a lot of Tegucigalpans are in here now. They kinda have taken over the role of all of the organizations, most of them . . . [and it] is actually a Tegucigalpan that is now a head of all them."[13] This change—some say—has furthered the gulf between islanders and organized conservation efforts. In the past, when Mr. Shelby was involved, many community members felt that BICA worked closely with the community, listening to members' needs and adjusting its programming accordingly. As Marta says, "What helped BICA back then was it was all about Shelby. It was Shelby and Glenn Pederson who started it and everybody knew it. Now it is being run by foreigners." Marta further outlined the issue as people from the Global North and

Figure 14 Exhumed turtles in 2016. Notice the two cameras taking photos as the BICA staff member exhumes the nest. My camera makes a third, and at least three other tourists were present with their cameras. Photo by author.

the capital city of Tegucigalpa are living with and experiencing the impact of anthropogenic environmental change and thus feel compelled to "solve" problems in places such as Utila. Utila is a small place, she argued, and while the problems may be severe, they seem manageable. "Places in the developed world are already a lost cause," Marta feels. Marta is articulating Moore's (2019) point that island places often serve as Anthropocene laboratories, places for global change science studies, experiments, and interventions, run by conservationists from elsewhere.

Environmentalists from elsewhere are locally understood to have a very different subjectivity and sensibility than islanders. A white male islander, comparing BICA staff from its inception to that in the 2010s, shared his views on professionally trained scientists from the capital coming to the island to run the organization. "They think a lot different than we Islanders. They do think different. They're a little kinda selfish. Um, kinda egotistical and maybe a little greedy and not caring too much about the Islanders and stuff like that." Back in 2016, this man shared that he hears islanders claim that BICA is not doing anything, that they are "only trying to make money now . . . trying to bring in money and stuff for projects." I propose that this perceived transition in community engagement is not altogether a result of a difference in subjectivities and the influx of selfish scientists from the mainland but rather the fact that neoliberal conservation wears away at the organization's effectiveness at a local level.

When environmental governance decentralized from being state funded to managed through NGOs with limited to no public funding support, NGOs became solely responsible for generating financial resources for their programs and initiatives. This predicament leads most to operate at the mercy of grant and foundation priorities, which may or may not closely align with local needs. As BICA and other organizations work toward financial stability, they find themselves managing projects that were brainstorms from other people in distant places (including partnerships with me). They also find themselves moving much more squarely into selling experiences with nature through conservation voluntourism, a talent first developed on the island with the Iguana Station. All of this fundraising serves to detract time and resources from doing the scientific studies and community programming that the organization was established for. Along the way, the local community loses employment opportunities, engagement, and confidence in the organizations.

Soul-Searching While Saving the Swamper at the Iguana Research and Breeding Station

The second organization to emerge on Utila was the Iguana Research and Breeding Station (often simply called "the Station" by volunteers). Staff and volunteers at the Iguana Station work to preserve the endemic "swamper," or "wishiwilly" (*Ctenosaura bakeri*) and its habitat, as well as promote sustainable development on the island through environmental education programs. The facility began as a German-financed scientific project in 1994, with a facility constructed in 1997. It was turned over to the Hondurans in 2007 and is now run by the Fundación Islas de la Bahía (FIB). Like all NGOs on the island, it also struggles to maintain financial solvency. Fees paid by volunteers serve as an instrumental source of income to keep the station afloat (but yet they are just barely afloat). As with BICA, leadership changes hands frequently. There have been five different directors since I started working with the Station in 2011. Most have come up into the director position upon completion of their own practicum work for a degree in biological sciences from a Honduran institution.

In 2011, I participated as a full-time volunteer with the Iguana Station (since then, all of my other participant observation with volunteer programs on the island have been one-offs, on a day-to-day or activity basis). I offer observations from the 2011 Iguana Station experience to provide a glimpse into the emergence of the conservation volunteer subjectivity and offer the following snapshot of the volunteers I engaged at that time as window into the motivations of conservation volunteers and their core takeaways about the island, its inhabitants, and their perception of the main ecological threats facing the island.

The majority of volunteers I encountered in 2011 were young people on a break from university studies, typically from the UK. All but two ($n = 15$) had arranged their trips through an umbrella organization that coordinates volunteer holidays such as Original Volunteers (Volunteer Forever n.d.). Only one volunteer was Honduran, and this volunteer, like all nationals, was not charged a fee to volunteer. Utila was typically one of two or three "stops" they would make that summer, most continuing on to volunteer programs in additional Central American countries (usually Costa Rica). Volunteers stayed approximately four to eight weeks on Utila (the Iguana Station required a minimum stay of three weeks).[14]

Volunteers rise early and work at a leisurely pace from 8 a.m. to 12 p.m., taking a two-hour lunch break, and then resuming work for three more hours (2 p.m.–5 p.m.) in the afternoon. Thus, volunteers work a minimum of seven hours a day, though some choose to work more. Activities include feeding and caring for captive iguanas, housekeeping of the volunteer barracks, groundskeeping, providing tours at the visitor's center, collecting crabs and termites for iguana feedings, providing environmental education workshops for local schoolchildren (at the Iguana Station and at schools), tracking and documenting iguanas in the wild, and providing research assistance to staff scientists. Volunteers work out their daily assignments among themselves in a weekly meeting with the director and volunteer coordinator. At these meetings all weekly "trips" are announced, and the volunteers can decide if they wish to go on any. These trips are half-day journeys into the mangroves and can be quite physically demanding. Not only is navigating through the mangroves an extremely arduous task because red mangroves grow on aerial prop roots in arches above the water level, but just getting to the mangrove forests from the station involves a few-mile walk across the hilly portion of Utila, usually in the summer heat.

In my time at the Iguana Station I observed fairly extreme isolation of voluntourists. Seldom did they leave the complex, which is situated up a long hill, fairly well removed from the main tourist strip. When they did leave during the day, it was for trips to the mangroves, a place where they rarely encountered other individuals. On days off or in evenings, the volunteers joined the normal tourist crowds in main street restaurants, bars, and beaches, most frequently visiting with others similar to them. Volunteers largely spoke only to the Utilians employed by the Iguana Station as groundskeepers or iguana trackers, with teachers and students in the local schools, or with island environmentalists.

Like turtle, iguana meat is a beloved traditional dish, consumed by generations of Utilians ever since the first settlers arrived on the Cays. Islanders have fond memories of the past when hunting iguana was not illegal, celebrated rather than frowned upon. Iris, a phenotypically brown but white-identifying islander, shares, "Everyone had dogs around the house, so they go hunting iguana. It was just a fun thing to go run the iguanas. Because with the dogs running, you would be getting the iguana out of the tree and the dogs would be down below, and then the dog runs and you runs behind the dog so the dog won't eat the iguanas. So that was fun days for us!"

Most Iguana Station staff were aware that their mission ran counter to local culinary traditions, and as a result they did not focus their core messaging on changing local tastes. Rather, they understood that the primary threat to the swampers was the incessant clearing of their mangrove habitat. However, this nuance was almost completely lost on the volunteers themselves, who repeatedly stated that the biggest threat to the swamper was the islanders eating them. In all of my interviews with volunteers, eating iguana was mentioned as one of the key environmental problems on the island. And when I asked volunteers to describe the local Utilian relationships to the environment, I received responses such as the following:

They aren't very concerned about the environment. It's not important to them—they throw trash, eat the iguanas, don't care for the sea. They aren't worried about the environment. (twenty-year old Honduran, recent university graduate who has been volunteering on and off for years at the Iguana Station)

This island's got so much potential. I mean it's amazing. But if the beaches were a bit cleaner or people were more about [conserving] the animals. If people knew more about the animals and how to conserve them and how to live better with them I guess. 'Cause I didn't realize for a long time that people ate iguanas here. I mean I think, this is the only place where you have them. Why would you eat them? I mean this is amazing, just to have something that no one else has. (twenty-seven-year-old British nurse on a three-month career break)

I don't know how interested they are in this kind of stuff. I mean they eat them. I'm worried that they think we're—not like imposing—but just bringing our ways and taking over. Because if these people got no money and they are eating these iguanas; I mean, if I had to feed my family and I didn't have the money, I'd do anything I had to. So I worry that they think that "you're coming here; you're taking like—you're stopping our livelihoods; you're making things difficult for us." And maybe they need to be better educated about iguanas. Because people were saying to me that the iguanas die, the swamps go bad, the sea goes bad. Maybe if they knew more about that, they'd understand why we're doing what we're doing. (twenty-one-year-old British female student)

Volunteers often concluded that locals did not possess the same knowledge that they did about the importance of native species protection. Further, the consumption of iguana meat was taken as an indicator that locals were generally disinterested in conservation and did not care about their environment.

Volunteer conclusions that locals did not understand their environment emerge because in the context of conservation voluntourism, nature is commodified and relationships to local resources are obscured. To create nature as a true commodity, people and local livelihoods dependent on specific lands, waters, and natural resources, have to be detangled and removed. Or, as J. Igoe (2013) and S. Sullivan (2013) argue, in order to "make nature move," it first has to be made to "sit still as an increasingly deadened object of contemplation" (Igoe, 2013). On Utila, "to make nature move," the endemic swamper and its mangrove habitat had to be transformed into commodities for "protection." This form of capital circulation privileges a new form of labor from that of the past, which was a labor of hunting iguana with one's dogs and then cooking up a traditional stew with the meat for one's family.

The labor that becomes valued in this new system is now associated with protection of specific species and ecosystems, and in particular activities that work to produce a view of nature as a place void of local hunters (now labeled "poachers") or of local fishers (now seen as illegally sourcing in protected zones). The newly valued labor is one of affect and care, associated with conservation researchers and volunteers who engage in activities associated with species protection. With nature the commodity and a "spectacle" that privileges a view of nature void of human activity (unless it is an activity coded as "science"), volunteers present for short-term work lack both micro- and macrolevel understandings of resource management and conservation. They are left only to focus on conserving resources *away from* local peoples.

In focusing on conserving species without local consultation, volunteers miss critical knowledge that locals possess, which emerges out of intimate knowledge acquired through generations of experience in the bush and a relational form of worlding. One area that Utilians track without western science is with respect to species reproduction rates and emerging threats to survival. Islanders may consume iguana, but they also have important perspectives that could inform community-based management plans, based on their understanding of iguana reproduction and ensuing threats to their survival.

I wish to take a detour back to crabs, another favorite local dish, and crabbing, a livelihood strategy for some, and enjoyable activity for others who like to source their evening meal from bush to table. First off, remember that iguanas eat crabs, crabs eat mangrove refuse, iguanas dwell in mangroves, and people eat both crabs and iguana. All of these species assemble together on Utila and rely on another. This way of worlding is relational, interwoven, and entangled, a stark departure from the one-world-world framework.

Crab hunters—who go out at night when the crabs are plentiful (or at least when they had been plentiful in the past)—have begun to encounter a growing number of raccoons on the island. The most common origin story I heard was that about ten years ago, an island resident with aspirations of starting a zoo acquired a handful of raccoons. A handful of years ago, he determined a zoo was not going to work out, and so he set them free in the bush, a seemingly benign thing to do. Yet, raccoons are not endemic to the island and they have since multiplied to such an extreme extent that they are wiping out a large percentage of the crab and iguana population, not to mention the impact they are having on the island's fruit trees, birds, snakes, and insects and other creatures.[15] Islanders are keenly aware that this introduced species is threatening their livelihoods and preferred meals. Some crabbers have taken to killing them whenever they can.[16]

We return now to the iguana for another example of islanders' knowledge of the reproductive rates of local species, without western science telling them. Islanders are well aware that the nonendangered *Ctenosaura simili* (black spiny-tailed iguana, known as the "highlander" by locals) outreproduces the swamper. Seeing a plentitude of baby highlanders hatching in one's backyard suggests to many islanders that these nonendangered iguanas should not fall under protected status (it has been illegal to hunt iguana in Honduras since 1994). In 2019, walking through the yard of a white islander Old Head, Mr. Richard, with a group of young adults in BICA's Reef Leader program, he tells us about the various iguana he sees in his yard. He shares that he sees swampers all the time, because his grounds are full of black mangroves. He slaps a tree and says, "The swampers are evolved with the black mangrove tree. Black mangrove tree—the body's is black and so the swamper evolved with it. They usually have holes in it, and that's where the swamper goes to hide." Our team asks him to tell the difference between the various iguanas on the island. He tells us a bit about the green iguana, about which he says, "sometimes the male comes out red," and then moves on to describe the highlander and to suggest that the current protected species laws are out of sync with the reality of the island.

> The highlander is brown with black stripes. Very prolific. They lay like eighty eggs. They eat your baby chickens worse than a snake. Worse than a boa.

A highlander [*laughs*] you know, we're protecting all the stuff like that, but whooo! I think you can leave 'em so people can hunt the highlander. The swamper [*emphasizing his words*] *should not be hunt at all* because they only lay 5 to 25 eggs. The highlander lays up to 80 eggs. They shouldn't be on the list for preservation because they are [*snapping fingers to emphasize their laying of eggs*] so prolific; they are so plentiful. [*pointing to his grounds*] They're coming for eggs right over there in that soil, hatch a whole bunch of them off. So highlanders could be on the side. You could let people eat a little bit. You can't change habits of everybody like that. That's how we kinda get messed up with our conservation and stuff like that. Because you prohibit everything, and people resent that because they are used to eating swamper.

Mr. Richard's suggestion of a culturally appropriate, community-based management iguana policy might not be as far-fetched of an idea to conservation professionals working on the environment, yet the eternal problem facing the Iguana Station and their counterparts is one that often besets small NGOs under neoliberalism. As an underresourced and understaffed organization, the focus of the Iguana Station's limited staff has to remain centered almost entirely on simply keeping the Station afloat. In 2011, there were only two regular staff people, one the full-time director and the other a part-time volunteer coordinator (who typically worked full-time hours anyway, out of the need for additional support). Others were hired on a temporary basis, as iguana trackers or groundskeepers. The research station relied on paying volunteers to keep their operation in business. Volunteers ended up doing the day-to-day chores that, if funding were available, a full-time employee would be performing. And with just one and a half employees (the director and the volunteer coordinator), very little time was available for actual "research," let alone collaboration with the local population.[17]

THE TWENTY-FIRST-CENTURY NEOLIBERAL CITIZEN

The case of volunteerism at the Iguana Station illuminates how a hierarchy of value has emerged through the "spectacle of saving" associated with conservation voluntourism. Rather than pay local people for the labor involved in caring for their natural resources or valuing the work they already do to steward their surroundings, international tourists pay to experience protected areas and resources (reefs, mangroves, swampers). Reliant on external sources to keep themselves afloat, the growing voluntourism sector offers another revenue source to the unpredictability of granting programs. This sector continues to grow as global citizens work to accumulate the cultural capital and competencies that are valued under twenty-first-century neoliberalism.

The value these voluntourists gain is not economic value for themselves in the moment. To the contrary, they are expending economic resources, but they are gaining nonmonetized value and experiences that hold the potential to be translated into the creation of economic capital in the future. If the labor of volunteers might, in a different context, be afforded an economic value, why were they doing it? What motivated them to travel to Honduras to clean barracks, cut up hibiscus flowers to feed baby iguana, climb inside cages to clean iguana droppings, or capture crabs in black mangroves to feed the adults? Volunteers at the Iguana Station reflected these patterns. The following quotes are illustrative.

> I thought I'd come here to learn something about myself... I kind of wanted to be able to prove to myself that I could handle myself and get to a new place, and I'd be able to communicate with the right people to get where I'm going. And that was a part of planning the journey, 'cause I had to find out all about it myself. I kind of proved to myself that I can do these things. (twenty-one-year-old male college student from England, majoring in environmental science)

> When I finished the university, I was looking to do something because I didn't have work and I like to help, so I came here. I wanted to do something that would help with my career. (twenty-seven-year-old Honduran female volunteer from Honduras' capital city)

These quotes capture the spirit of the volunteers I encountered, all of whom were largely on a "journey of discovery." Even those who expressed a desire to "help," such as the second woman quoted above, coupled their altruism with a motivation to develop their own competencies. Through the volunteer experience, individuals are gaining the "entrepreneurial" competencies necessary to compete in the new neoliberal economy. In her critical analysis of voluntourism, Wanda Vrasti (2012) advances Foucault's ideas from *The Birth of Biopolitics* (2008) to argue that *Homo economicus* no longer fully applies under twenty-first-century neoliberalism. *Homo economicus* urged individual actors to assume responsibility for their own actions and treat everyone else as though they too make decisions based on cost-benefit logic. However, Vrasti points out that market rationality is no longer enough. Today's neoliberal subject must complement their economic rationality with an "entrepreneurial spirit." Vrasti writes (2012, 21),

> The new entrepreneur... is not asked to dispense with economic rationality, only to complement it with what were once bohemian and counter-cultural dispositions. Instead of the rational, calculating and cold-blooded *American Psycho*, the good neoliberal subject of the 21st century is the rather schizophrenic figure of the compassionate entrepreneur, the happy workaholic, the charitable CEO, the creative worker, the frugal consumer, and last but not least, the volunteer tourist.

Scholars find that many gap year programs are explicitly marketed to youth tourists in ways that suggest young people will gain experiences necessary to compete in a highly competitive economy (e.g., Heath 2007; Söderman and Snead 2008). Participants gain an "economy of experience" (Heath 2007) that provides a competitive edge over other job seekers. The qualities associated with independent travel are seen as particularly well suited to the post-Fordist economy and flexible employment (O'Reilly 2006, cited in Lyons et al. 2012).

By and large, volunteers told me they were looking to help others while gaining skills for their resumes.[18] Resume building is not the sole reason volunteers choose work with other species. Many are seeking reconnection to the natural world and a departure from their technology-saturated worlds (even if just for an instant while their hands are full and phones in their pockets). Most volunteers believed they would experience personal growth through their exchanges with other species and landscape assemblages. Some explicitly chose the Iguana Station because they wanted to challenge themselves to do things they would not likely have done otherwise, for instance, hold a tarantula or trek through mangroves to capture iguanas.

Through these experiences, their entrepreneurial competencies are validated and enhanced, their desires to connect with other species satiated, and their souls nourished by engaging in acts of care, and they return home to their lives as fully constituted neoliberal citizens. Even if voluntourists do not interact with others, their presence in a different space enables them to return home with the credentials needed to succeed in becoming full neoliberal citizens; they have earned the social and cultural capital and entrepreneurial competencies that enable mobility and "success" within the Global North. Vrasti (2012, 23) writes, "In dedicating their time and money to helping the global poor, volunteers display precisely the types of qualities needed to assume a privileged subjectivity: an ability to operate in distant and diverse settings, a desire for social change and an interest in experimenting with one's self and the world around it."

While volunteers are expanding their resumes, the places and peoples that host these charitable endeavors are excluded from the possibilities that the voluntourism experience generates. Most crucially, the new neoliberal economy appreciates and rewards those who can travel freely into spaces and assume activities that were once pursued (or could be pursued) by original inhabitants, denying that territorial belonging ever even mattered (Vrasti 2012). The cultural competencies and forms of capital that are valued within this emerging economy of affect accelerate dispossession for Utila-born islanders.

"Save a whale, drink a beer" at the Whale Shark Oceanic Research Center (WSORC)

Posters at the WSORC fundraiser for 2013 International Whale Shark Day encouraged tourists to "feed me your lempiras" and to "save a whale, drink a beer!" (http://wsorc.org/international-whale-shark-day-celebrations/). Contrary to what this slogan suggests, whale sharks are fish (sharks), not whales.

The Whale Shark Oceanic Research Center is a marine-based private organization established in 1997 with a mission to "increase the presence and availability of researchers in Honduras that work with whale sharks." WSORC holds the only permit in Honduras to study whale sharks, and its staff were responsible for developing the legal guidelines for whale shark encounters, approved by the Honduran government in 2008. In addition to hosting conservation voluntourists who help collect photographs of spot patterning for Wildbook, the international NGO that collates global whale sightings such that users can track movements (whale shark spots are unique, similar to a human fingerprint), WSORC also runs "Ocean Safaris" to take tourists out for responsible whale shark encounters. WSORC operates these excursions through the Bay Islands College of Diving (BICD) and Utila Lodge; all three (WSORC, BICD, and Utila Lodge) were founded by the late Jim Engels, an expat from the United States who served as an early member of BICA. BICD / Utila Lodge is one of the biggest dive resorts on the island.

The founder of these joint operations has capitalized on the increasing entanglement of protected species research and tourist desires. Take this 2017 advertisement for a five-night all-inclusive package from its affiliated Utila Lodge and dive shop as an example (Utila Lodge 2017):

WHALE SHARK ENCOUNTERS

It is the reason that many of our guests come to Utila—to swim with whale sharks! The gentle whale shark is a filter feeder and can grow to be over 12 meters in length—the largest fish in the ocean!

Utila is one of the very few places worldwide where whale sharks can be reliably found year round. With years of experience at finding the elusive sharks, our captains will do their best to make sure you get a (or several) good looks at these magnificent animals. Most mornings, weather permitting, our dive boat ventures to the north side of the island which borders the Cayman trench. It is thought that the up-welling currents from the deep may provide nutrients that attract the small ocean creatures upon which whale sharks feed. Once a whale shark is found, our guests get the chance of a lifetime to jump in and snorkel with a whale shark!

On top of promised whale shark encounters, this $985 all-inclusive package covers meals, accommodation, three daily boat dives (including night dives), kayaking, and even a welcome bottle of wine. WSORC's daily Ocean Safaris take tourists to the north side of the island in search of whale sharks, dolphins, tuna boils, and rays. Beyond standard tourist packages, WSORC runs a formal internship program that attracts university students from around the world (though the vast majority come from the United States). Internships range from one to three months and range in price depending on the additional diving courses a student wishes to undertake. At approximately $400–$500 a week, the programs are the most expensive of conservation internships or conversation voluntourism on the island, but still considered fairly affordable by U.S. standards. WSORC internships require scuba certification and thus provide a nice opportunity for BICD, which is on the same property, to collaborate on instruction. Interns are trained in coral and mangrove nursery management, lionfish containment, and whale shark monitoring. WSORC has partnered on a variety of projects with other conservation groups in the past, including a lionfish ecology program that was being run by Operation Wallacea. WSORC also collaborates with the other organizations on the K–12 environmental education program, coordinated by BICA.

WSORC typically recruits its leadership team, and thus conservation science instructors, from past interns. For example, the 2019 program manager and internship coordinator first came to Utila in 2016 on a one-month internship after completing her BA in English literature and film studies. According to her public profile, like many conservation voluntourists and interns, she "delayed her flight home," staying on the island to continue dive training at BICD. She then became a dive instructor at BICD and spent two years there before moving to WSORC. WSORC's 2019 research coordinator was recruited from Operation Wallacea after serving as a site manager in Utila and Roatán for the past three summers. The critique that WSORC relies almost entirely on foreign volunteers and does little to promote conservation careers for locals and mainland Hondurans seems to have been heard by the organization's Board of Directors, who in 2019 launched a scholarship in the name of the late Jim Engels (the Jim Engels Ocean Steward Scholarship, or JEOS). This scholarship supports a six-month internship with the organization and covers housing, dive training, and marine conservation coursework; food and training materials are not included. In 2019, WSORC had brought on a former JEOS recipient from the mainland to serve as its community outreach coordinator.

As with the other small conservation organizations, WSORC leadership is constantly changing, and even more often than FIB and BICA. None of these three individuals I just described were in these positions in 2018. As well, the lead coordinator is almost always a US citizen who came to Utila through the internship program. The staff are often in their early twenties and between academic programs and permanent, long-term employment with benefits, likely contributing to the very high turnover. As a longer-term conservationist from another organization on the island noted, "There's always a new one. Just replace one blond woman in a whale shark suit with another" (see figure 15).

WSORC's most developed page on its website is its internship page, which lists opportunities ranging from a six-day citizen science program to a four-

Figure 15 Screenshot of WSORC social media. The "whale shark" wetsuit makes its rounds among interns. wsorc_life Instagram. Screenshot taken by author.

week marine conservation internship, to dive master and instructor training. The latter require a minimum commitment of two months and are priced at just under $2000/month. These programs serve as a core recruitment tool for future WSORC internship coordinators. To some, WSORC might be understood as first and foremost a tourist facility with some lighter science occurring on the side. As far as I know, it currently does not have researchers on staff who are focused on large-scale whale shark science, instead relying on training up volunteers to deliver lectures.[19]

◆ ◆ ◆

Tourism imaginaries play a central role in shaping the ways in which various actors in the conservation voluntourism industry assemble and the resultant emergent subjectivities. Following Noel Salazar and Nelson Graburn (2014, 1), imaginaries in this context refers to "socially transmitted representational assemblages that interact with people's personal imaginings and that are used as meaning-making and world-shaping devices" (see also Salazar 2012). Imaginaries are implicit interpretations that are often structured in dichotomies (e.g., host-guest; here-there; nature-culture; local-global). Capturing and packaging these dichotomies into a tourism product enables the tourist to move from imagining to living that imaginary. Because imaginaries are implicit and intangible, they can only be studied through the ways in which they become visible and tangible in circulated images and discourse (Salazar and Graburn 2014, 1–2). One can begin to infer the meanings and impacts of tourism imaginaries from what people say and do and from what they choose to memorialize and circulate. Examples of making tourism imaginaries tangible range from museum exhibits and cultural performances to hotel or airport art to tourist Instagram pages.

Tourist desires and voluntourist experiences are influenced both by what they know and what they wish to demonstrate to others that they know, and what they know is influenced by the spectacle of nature associated with tourism imaginaries. Take, for example, the following blog entry by an intern/conservation voluntourist at WSORC; here she reflects on her experience with mainstream tourists, as they were both on excursions to encounter whale sharks, albeit with different hosts (Van Landeghem 2017, n.p.).

> When the boat sailed out the excitement only grew. Looking at the horizon for birds or a sign for a tuna, we waited for those magical words "There's a whale shark!" Then it was time to get our gear on, waiting in tension for the "GO GO GO" of our captain. Getting into the water as calm as possible, seeing that majestic creature eating, swimming underneath you and eventually watching him disappear into the big blue ocean. In silence we go back to the boat, then our excitement comes out . . . It's still the most incredible thing ever!
>
> Limited amount of people with the animal, no in-water noise, no touching, keeping distance, no stress for the animal & maximal encounter time.
>
> This is how it should be.
>
> This is how it could be.
>
> But unfortunately this is not reality.
>
> Up to 4 other boats came up close. Doesn't matter how many people are already in the water, every one on those boats jumps in, splashing & screaming around the animal, duck-diving to get that one 100-likes-Facebook-selfie, boats blowing their horns or even cruising in between the snorkelers. And this story repeated itself several times.
>
> Whale shark gone.

Boil gone.
Magic gone.
I'm done.

The title of this entry is "when you're aware, you care." The entry does not end with her being "done." Rather, the author's experiences invoked in her a sense of responsibility to blog and communicate publicly the rationale behind whale shark encounter regulations. Her post documents the lived reality for whale sharks as they indeed endure stress by the dozens of loud tourists in close proximity, who are trying to get photographs (and sometimes even photos to appear as if they were riding a whale). Tour companies on the island feature videos on their Facebook accounts promoting whale shark trips with "#TheGentleGiant." For instance, one of the major tour guide's Facebook page features videos where you can hear tourists asking once the boat has sidled up to the whale sharks if they can go in, someone then saying, "Go ahead! Jump on in. Slide on in. He ain't going anywhere." The camera then lands on a man snorkeling directly on top of the whale shark, with several other tourists surrounding nearby creatures.

The blog author likely encountered a tour of this nature. She likens the experience to being a "superstar, eating at a restaurant, [when] suddenly 15 screaming paparazzi come to your table." Of course, you'd take off, and quite possibly never come back to that restaurant, she notes. Hoping to spread awareness, this conservation voluntourist explains that by learning and abiding by the regulatory code, informed tourists can engage in multi-species encounters marked by care. And, future conservation voluntourists at WSORC can translate their affective labor into scientific data collection, taking photographs to add to the Wildbook database.

While the emergence of whale shark tourism and the conservation science connected to it contributes to global conservation initiatives, it has also shifted relationships to Old Tom for Utila's fishers and boat captains. In speaking to fishermen in the Cays in 2020 about the decline in the fisheries, Mr. Frank, an older white settler descendant, shared, "I think it's all the whale sharking. All the tourists are scaring 'em off and they the ones who brought the tuna."

Another outcome of the rise in whale shark tourism is that tourists (both regular and conservation volunteers) now make decisions about whom to hire for a whale shark encounter based on endorsements of conservation voluntourists who engaged with the private organization that produced the regulatory code (WSORC). It further serves to note that the guidelines for responsible encounters that have been produced by foreign conservation voluntourists have taken shape in the format of an industry that encourages in-water multispecies encounters, a sharp turn from the fisherman's appreciation of Old Tom's presence from inside a boat.

Replacing earlier multispecies assemblages of Old Tom, bonito, and fisherman, with whale shark tourists, conservation scientists, and boat captains, has material and affective effects. The Utilian fishermen had (and continue to have) a relationship to, and knowledge of, Old Tom by living so closely among the sharks, fishing daily in an economy that is affective (they know and care for Old Tom's well-being) and material (Old Tom helped fishers procure food). The emerging affect economy under conservation voluntourism produces different multispecies assemblages and material outcomes. Conservation tourists are purchasing cultural capital through an experience of caring for the whale shark, which they will later exchange for material outcomes. One might also argue that fishers were and are more in sync with the whale shark—they know the rhythm of the whale shark; they respect its presence as connected to a wider assemblage of beings, including bonito, plankton, corals, and other marine life, something lost to those focused on snapping selfies.

Killing Lionfish to Save the Seas

> Honestly, nothing beats killing a lionfish. It's such a thrill! I think you guys will get a such a kick out of it.
>
> —Volunteer trainer, introduction to culling program

Three organizations partner on lionfish culling and lionfish ecology studies on Utila: WSORC, BICA, and Opwall.[20] Opwall brings paying conservation voluntourists to assist with dissections for stomach content analyses and population studies, but they do not allow their volunteers to spear lionfish.[21] Conservation volunteers and dive tourists who undergo BICA training to obtain a permit can engage in culling with a Hawaiian sling. Lionfish culling has become part of Utila's protected area management suite of strategies, and only those certified through the training can participate in hunts. There are only a handful of local fishers who are licensed to use the Hawaiian sling, and thus the organizations work collaboratively with Utila's dive centers to train interested dive tourists, who end up playing a key role in lionfish hunts. In practice this means unlicensed Utilian fishers could be fined for spearfishing lionfish (because all forms of spearfishing are illegal in the Bay Islands) while volunteers who can pay for the training and permit can participate in culling and lionfish derbies. Thus, the organizations collaborate to hunt and kill lionfish, which are then sold in local restaurants to tourists. This program is one of just a handful that brings some unrestricted funding to the financially struggling conservation organizations.

Marine ecologists studying mesophotic reefs suggest that this management approach may not be effective, largely due our inability to appreciate the life cycle of the lionfish. Mesophotic coral ecosystems are found at depths ranging from about 30 to 150 meters and are marked by communities of corals, sponges, and algae. As one marine ecologist shared,

> I came out here to work on mesophotic reefs, and everywhere I looked I saw lots and lots of lionfish in really deep reefs . . . Lionfish exist from really shallow reefs down to 150 to 250 meters, so in Utila if you snorkel in these shallow reefs you very rarely see lionfish because we've built such good culling programs. All the dive centers have people going out culling. As soon as you go below thirty meters they're everywhere; there's huge numbers of them . . . Also their life cycle is when they spawn they produce these eggs that then float to the surface; young lionfish then move to shallow habitats, such as mangroves, seagrass, and shallow back reefs, and then as they mature they migrate out to the reef crest, and then down the slope as they get more mature . . . Are we leaving this big lionfish refuge down there that are constantly spawning, and coming back up to the shallows? So, is the management plan really working? Or are we leaving this big population? Because in fish, the more mature they are, the more fecund they are . . . So not only are these deep lionfish being missed by the culling, but they're also, because of this life cycle, the most mature ones that are producing the most eggs.

The culling programs may not be producing the conservation goal they hope to in terms of long-term lionfish management, but they are producing value for both conservation volunteers and Utila's hosts. The lionfish-culling program creates data for conservation volunteers who are studying lionfish feeding preferences through dissection and gut content analysis. They also contribute to Utila's lionfish derbies, where island restaurants engage in a cook-off competition and party attended by tourists at least twice a year, sometimes more. The lionfish derbies, the culling programs, and the ecological science attached to island conservation organizations all produce value in Utila's affect economy, as a tourist attraction when voluntourists care for the survival of some species through the death of another. The program has also been instrumental in bringing BICA, Opwall, and WSORC closer together to collaborate on joint programming and data sharing. As well, the lionfish science and culling activities rationalize and legitimize the presence of tourist-visitors, who unlike fishermen families from Utila's Cays or migrants from the mainland are cast in the language of belonging. They are there to contribute to the economy and to manage an invasive species, whereas the latter, especially migrants from the mainland, are more frequently subject to claims of overexploitation of species and lack of belonging (through the language of invasion, much like the lionfish). Moore (2012, 679) found the same situation in her study of lionfish and fisheries management in the Bahamas: "This figure of the invasive arrival, into which the lionfish fits all too nicely, is contrasted with that of the overly welcome visitor, the tourist and offshore finance investor who bring resources with them to be captured." As in The Bahamas, the lionfish in Utila stands to become something of an "emergent keystone species," which will then legitimate more conservation science research, invasive species management work, and voluntourism opportunities (Moore 2012, 679). Spectacle plays a role in advancing this narrative and this industry. It also shapes the contours of multispecies encounters that volunteers will have with lionfish.

If I ever had a doubt, the attraction was made crystal clear during an all-day volunteer orientation I attended in 2017 when the room of twenty-something young women erupted in excited giggles and a chorus of cheerful commentary whenever the topic of lionfish training came up. One young woman from the United States who frequently interrupted the volunteer training sessions to talk about her passion for marine life exclaimed, "I want to get a long string [to hang my catch] and be able to hold them all out [she poses how she would for a photo] and say 'yeah! I got these!'" This woman already knew how she was going to pose in photos to share her multispecies encounters through social media, celebrating her love for the sea through the care of killing.

As the lionfish-culling programs create new opportunities for the island's economy, they also bring other potentially harmful changes with them. For one, lionfish hunting by dive tourists and conservation volunteers changes the multispecies dive encounter from passive to active, creating opportunities for instrument mishandling resulting in speared corals (and this is indeed why Opwall does not allow their voluntourists to spear hunt). Further, while several parties are capitalizing on the affective labor of volunteer tourists who experience life through death, caring for some species while killing others, these practices dispossess some people (particularly unregistered fishers) of multispecies relationships in favor of others.

◆ ◆ ◆

Conservation volunteers, dive tourists, hotel and dive shop operators, and conservation organizations all assemble together in ways that have impacted the lives of fishers and refashioned their relationships to the landscape and resident species. In particular, social media is accelerating dispossession but through a savior discourse: "saving the seas," or as we will see in the next example, sometimes islanders are being saved from themselves.

"Saving Lobster Divers": The Health Impacts of Dispossessed Lobster Divers

The late Jim Engels of WSORC facilitated the establishment of Utila's only hyperbaric chamber, housed within the Bay Islands College of Diving, and worked with BICA to introduce a reef fee,[22] a $1 tax on divers that would be used to support three initial projects: (1) the aforementioned buoy project, (2) coordinated garbage collection to combat the rising waste associated with tourism development, and (3) the installation of the hyperbaric chamber and clinic to treat divers when someone gets the bends.

"The bends" refers to decompression sickness (DCS), or caisson disease, which occurs in scuba divers, astronauts, aviators, and others who work or play in compressed-air environments. For divers, it occurs when nitrogen bubbles form in the body's tissues and bloodstream, leading to pain, numbness, nausea, and paralysis. As a diver descends, the nitrogen in the air tank increases in pressure, and as the pressure increases, nitrogen is dissolved into the body's tissues. As long as the diver remains at pressure, breathing in gas in proportion to the surrounding pressure, the nitrogen will not pose a health risk. However, if a diver returns to the surface too quickly, or does not properly equalize their pressure throughout the dive, nitrogen builds up in the body's tissues, and excess nitrogen will come out as gas bubbles, blocking blood flow and disrupting vessels and nerves, producing DCS. The most common area affected in divers with DCS is the spinal cord, and classic symptoms include "heavy legs," paralysis or numbness, difficulty with balance and walking, dizziness, and confusion. Decompression sickness can be treated in high-pressure, oxygen-rich chambers (e.g., Utila's hyperbaric chamber). Most treatments involve multiple sessions over several days. In severe cases, someone with DCS may require six hours per treatment for more than twenty days. Utila's hyperbaric chamber requires four people to run it. Prior to the installation of Utila's hyperbaric chamber, anyone impacted by the bends had to be immediately flown to Roatán for treatment in the decompression chamber at Anthony Keys Resort.

Lack of training in regulating one's pressure, faulty equipment, and time pressure on local lobster divers, who are paid by the catch and therefore incentivized to return to the surface quickly to swap tanks, are the main contributors to DCS in Utila. Watching lobster divers on Utila Cays come in from a day of work is devastating. I do not have data on the total number of Utilian divers who have been permanently impacted by DCS, but I can report observing several—especially those who have done the job for several years—with the inability to walk without a limp. The impact is particularly acute at the end of a workday, as men come to shore to sell their day's catch (for a total of a couple dollars a day), stumbling onto the dock with a glassy, dazed look in their eyes.

In 2018, the rates of injury and death among Honduras' lobster divers along the Miskito coast received global media attention through a series of news stories in major media outlets. The spotlight turned local attention to Utila's own plight with DCS when two islanders died as a result of faulty tanks and one man lost his leg. In response, Kisty Engels (Utila Lodge / WSORC, and the widow of Jim Engels) and a foreign-born longtime resident in the Cays launched a GoFundMe campaign called "Saving Lobster Divers."[23]

The campaign organizers shared videos and photo footage of the 2018 incidents, along with a note that said, "We hope this article has grabbed your attention because the community needs your help." After more than two decades of history collaborating with the Utila Hyperbaric Chamber, the organizers observed "the necessity of having to treat lobster diver patients. As these divers have no way to pay for the treatments, the costs associated put a huge strain on everyone's resources." They realized the key cause of so many lobster diving-related injuries is "the lack of education around proper diving techniques." The campaign launched with a message that the organizers "needed to figure out a way to educate the divers, along with a way to keep them from taking advantage of the program." According to the campaign website, the first phase of the project was "to train 20 divers in safer ways of diving and teaching them how to hunt for lionfish and how to provide emergency first aid[;] the funds will also go to making sure the divers are outfitted with safe and working equipment and to contributing to a documentary being shot to illustrate the issues that the workers of this industry face" (Utila Lobster Diver Program 2018).

In the public write-up of the project, the organizers note that the project was underfunded (they raised $19,031 of their $54,500 goal) and thus they were not able to purchase the twenty sets of scuba gear that they hoped to obtain (a complete scuba set costs approximately $1,000). While they did not purchase the gear, they did have five full sets of scuba gear donated to the program and three sets of regulators and buoyancy control devices (BCDs). Additional donations by PADI meant that all training material (videos, books, and certification cards) was supplied at no cost for participating divers in the open water certification class. All of the donated equipment was then housed at a local dive shop, where divers who completed the training program would be able to check out the gear for a daily fee of $1 per day, to be used to maintain the equipment over time.

The project ended with training a total of twenty-four divers. The goal was to train twenty divers, but a total of twenty-four turned up the day of the training, evidence of the desire for safe working conditions among Utilian divers. Of the handful of divers I have had the opportunity to talk with in the Cays, they were extremely grateful for the opportunity to participate in the training, and those who did not have the opportunity to take the December 2018 course were hopeful that the program would run again so that they could participate. Taking two weeks away from work is a major commitment; not diving means not earning money. That so many divers participated, and so many more wanted to, tells you they wish to be working under safe conditions. However, in the eyes of some Cayans, fishers and divers included, the program has not been altogether successful.

Lobster divers claimed that nobody came to follow up on how it was going, and the regulators were not actually in use. The divers desperately wanted to provide feedback to the sponsors from Utila and approached me as a conduit. The problem was that the divers were under the impression that they would be given new regulators upon completion of the course. However, instead, they later learned that the equipment would be retained in a local dive shop and the divers must go there on a daily basis and rent the regulators for twenty-five lempiras each time (approximately $1). The rental fee cutting into profit is one downside, but on top of that loss, divers stated that they could not rent the regulators until 7 a.m., three hours after most begin work (a typical diving workday is 4 a.m. to 6 p.m.), further reducing their daily income possibilities.

Cayans were also unclear as to how the $19,031 that was raised was spent, since the equipment, training materials, and presumably instructor time were all donated. There were likely costs associated with running the dive boats and other undonated equipment costs, and it is also possible the funds were used to offset treatment at the hyperbaric chamber, which can get quite costly.[24] Some in the diving community were resentful of how often lobster divers would come in seeking free treatment, and in fact one individual told me in August 2019 that they were "starting to crack down on the chamber" because the lobster divers seem to think they can "come and use it whenever they want, for free." Many of the lobster divers in the Cays are young migrants from the mainland (locally called "Spaniards") or poor islanders of color, another sign of occupational segregation and discrimination with the lowest rung of the socioeconomic ladder doing the most dangerous work.

This risky work had not always been associated with the poorest islanders. It was an early occupation of many, coming with somewhat less risk when the lobsters and conch were more plentiful, and the region had yet to be saturated by competing divers boating in from the mainland. But several of the older generation gave up the work after facing too many lost or permanently damaged lives in their close community. One Cayan told of losing a close friend's son who went too deep too fast, one day in a row too many. The sixteen-year-old returned to the dock in extreme pain. They carried him up to a porch to have him rest, but the young man kept crying out to lower down his body. They moved him down from the porch to the lower house, but he could find no relief, even after being lowered all the way to the ground. As he progressively got worse, they realized they had to rush him over to the hyperbaric chamber. They loaded him in a dory and sped to Utila. By the time they carried his body onto dock, the boy "vomited his lungs right onto the dock; his tongue hanging out of his mouth." The man who witnessed this story said it was the worst thing he had ever experienced in his life, and from that point on he swore off diving as a promise to his wife.

As the above account shows, even with the availability of the chamber, Utila's lobster divers face death on a daily basis or, at best, long-term health impacts if they continue in this line of work for too long. As the head of one of the local "fish factories" in the trailer to the documentary this group was producing as part of their GoFundMe campaign shared, "You know, sometimes when these guys gets the bends, you put them in the chamber, they get over it, but you still see that little damage." Miss Nanci, the head of the family-run fish factory, is the middle-woman in charge of sales on the coast. She has purchased a handful of regulators and equipment for local divers to use, but she is not trained in diving and equipment safety herself nor has the resources to supply and maintain equipment for all divers. This intervention that she is able to make is minimal at best.

Much of the "Saving Lobster Divers" GoFundMe campaign is documented on social media, the existence of which inspires new conservation volunteers to come to Utila and participate in WSORC programming because it appears to be making a positive impact on the local community. I did meet a WSORC intern who was inspired by this program and sought to learn more about its impact through their internship experience. They were surprised to learn that nobody had followed up with the divers by summer 2019. Yet, if we consider the lack of integration with the local community of voluntourists shown by the example of Iguana Station volunteers, the neglect to follow up is really not that surprising. And, if we think about the constant turnover in organizational leadership illustrated in all of the above accounts, it is hard to imagine how new staff keep up on what happened before their time, let alone are able to conduct evaluation research on past programming. In addition, as organizations move toward pursuing volunteer tourist dollars through the development of "citizen science" programs, the organization's work at creating a positive and enjoyable experience for the voluntourist begins outweighing other organizational priorities. The organization becomes financially reliant on the income raised through volunteer placements (and you want the voluntourists to also have fun so they share their experience on social media, and recommend the placement to future volunteers).

With the rise of value in experiences that demonstrate openness, flexibility, and adventure, the *idea* of conservation stays alive. Through the rise of the affect economy the conservation idea has transformed into something quite different than the attempt to minimize wasteful use or the destruction of wildlife. Today's conservation is a tourism imaginary that involves affective encounters, including care in stewardship science (e.g., coral reef cleaning, seagrass monitoring, or fish identification logs) and care for the survival of select species through body mutilation or modification (e.g., toe cuttings from iguanas, tag insertions in turtles) or through death of others (e.g., lionfish, orchid bee).

This transition in turn means that voluntourists shape and transform the local landscape and relationships to other species. Prioritizing whale shark encounters or conservation interns hunting lionfish, for instance, dispossesses locals of their own relationships to lobster diving and fishing with Old Tom. And this industry is surrounded by spectacle, impressive images and stories that attract future do-gooders to take part in doing their share to protect the world's natural resources and improve livelihoods of others. It is fueled by Instagram, Twitter, Facebook, and other social media outlets, inviting future volunteers who read the social media celebrations of all the good work that conservation scientists are doing.

As we have seen in this chapter, due to the inequalities in relation to caring (affect), multispecies conservation voluntourism is having concrete and tangible impacts on both local ecology and social relationships to other species. With the rise of protected species conservation, geophysical spaces are encapsulated and a science emerges to protect and allow the movement of some species (human and nonhuman) and subgroupings within species into bounded space, while keeping others out (e.g., whale shark tourists over local fishers or paying lionfish hunters over lobster divers; sea turtles over island turtlers; swampers over raccoons and hunters). These boundaries are closely associated with a form of worlding that separates humans from other forms of life, with humans filling the dominant role in a hierarchical system.

Contestations over boundaries are closely tied to discourses and actions of "protection." Sometimes these acts of protection are codified into law, resulting from volunteer-led conservation research, which then produce and require different multispecies interactions. For instance, when foreign volunteer researchers work closely with local NGOs and governing bodies to write the regulatory codes, including those for whale shark encounters and lionfish spearing, this then shapes volunteer tourist decisions over whom to hire locally to enable a whale shark encounter and alters diving experiences from passive to active, creating opportunities for instrument mishandling that result in speared corals. The protected status of *Ctenosaura bakeri* means that the capture and preparation of iguana stew has moved underground, replaced by researchers who are tracking, capturing, and then clipping toes or spines for genetic testing, probing genitals for sex identification, and tagging them for future monitoring, with Honduran nationals facilitating the research experience. The "object" of conservation voluntourism continues to shift, paralleling the creative destruction associated with capitalism. Joseph Schumpeter (1942) considered creative destruction to be an "essential fact about capitalism": the fact is capitalism necessitates a constant search for new products, objects, and innovations—new "sinks" for capital to replace outdated or "used up" ones. With multispecies voluntourism, volunteers and conservation science brokers must be on a constant search to find new and meaningful ways for volunteers to experience and engage in affective and caring relations with other species.

Several scholars working on affect have argued that the human body learns by doing, and is thus constantly making and remaking human subjectivities and socialities.[25] If as Singh (2013, 190) says, "affective labor transforms local subjectivities," watching the emergence of this affect economy one cannot help but to ask: What transformations emerge when the bulk of environmental stewardship comes from nonlocal volunteers?

Chapter 4

The Political Ecology of Multispecies Conservation Voluntourism, and Limitations to "Becoming With"

Political ecologists have paid little attention to conservation voluntourists to date, despite the fact that they are becoming increasingly present in global conservation. This book has presented an opportunity for readers to think critically about the opportunities and costs of participating in conservation volunteering and research tourism. Within these pages, I have presented conservation voluntourism as an approach to anthropogenic environmental change enmeshed in spectacle, grounded in affective labor, and producing new opportunities for "becoming with," through multispecies engagements. I have argued two interrelated points: (1) that vulnerable species are commodified in multispecies voluntourism encounters, dispossessing and obscuring local relationships; and (2) that the conservation voluntourism industry is both producing and produced by affect, which is then exchanged for material outcomes by global volunteers, who typically come from a western, educated class, and adhere to a one-world-world ontology heavily influenced by spectacle.

The emergence of a tourism-dependent economy and tourist-spectacle-dominated worlding practices have dispossessed islanders of their material holdings, livelihoods, and cultural traditions. By 2020, the vast majority of available prime real estate (including land, mangroves, sand, and coral rock) in Utila had been sold and developed for tourism or housing for lifestyle migrants. The final stretch to develop was the beachfront along Pumpkin

Hill, which received electricity and paved roads in 2018, and plots quickly began to sell. With most "good" land and tourist facilities now in the hands of foreign-born individuals, newer generations of Utilians are increasingly constrained in their housing and employment options. Much of the local employment opportunities have shifted from family-owned enterprises to service work for foreign-owned restaurants and dive shops. Once dispossessed of land and former agricultural and fish markets, settler families then faced a rhetorical dispossession as new values emerged through dive tourism, which privileged "protection" of endemic and endangered species that islanders had valued as food or as locaters for food (in the case of Old Tom).

The science of conservation and entwinement of conservation and tourism development led to a new hierarchy of ownership, labor, and care on the island. Today's affect economy privileges the external systems of science, conservation, and care, and it generates and exchanges multispecies experiences for new forms of capital in the global economy. Even if the conservation volunteering industry is underdeveloped and the actual numbers of voluntourists participating in these activities pale in comparison to the numbers of "regular" tourists who come for diving, snorkeling, deep-sea fishing, and whale shark encounters, the visual presence of these organizations on social media generates increased numbers of travelers to the island. The tourism imaginaries that are produced through images that emerge from the conservation organizations suggest that Utila holds opportunities for people and other beings to come together to exchange affect, to enter into multispecies kin relations, to become with, and to potentially imagine alternative ways of worlding.

Humans have always been entangled in multispecies assemblages. We have always been becoming with, but how we assemble and how we become with are shifting through the emergence of new affective economies. The economies of affect that are developing through multispecies conservation voluntourism are distinct from the economies of affect that previously constituted multispecies encounters on and around Utila. In earlier years, the affective component of socioecological assemblages was largely related to a close familiarity with various species that islanders relied on for material benefit (for sustenance or sale) as well as for their cultural importance as part of local dishes served during family gatherings and holidays. It was a worlding framework of relationality. Families hunted together to capture iguana or turtle for local dishes, and fishers drew on their understanding of Old

Tom's feeding patterns to help them locate catch. As we read in chapter 1, island children kept small turtles in their family crawl and once the baby turtle matured, it would be released into the sea. Islanders once enjoyed the "sweet delicious meat" of green turtles, and the care work associated with raising the babies can be understood as a practice that ensured the species survived into adulthood, enabling the eventual harvest of eggs or capture of occasional turtle for consumption. Today, harvesting turtle and their eggs is illegal, and thus nobody talks openly about eating them; they do, however, reminisce about the taste from their childhoods.

While the above multispecies assemblages continue to exist in part, with the growth of conservation and tourism development there are now new actors transforming how species assemble (e.g., migrants, tourists, volunteers, and lionfish). The shifting entanglements under the affect economy associated with conservation and nature tourism fetishize nature, reducing vulnerable species to commodities to serve the emotional needs of tourists and volunteers. Just as is the case of humanitarian volunteerism in which the images produced by humanitarians focus on a suffering subject to generate compassion and justify intervention, commodifying a particular category of people (AIDS orphans, hurricane victims, etc.), conservation voluntourism rests on the sanitization and depoliticization of other forms of sentient beings (Coghlan 2007; Freidus 2010, 2017; Vodopivec and Jaffee 2011). Utilians, migrants, ecologists and biologists, voluntourists, and nature tourists now assemble through exchanges of affect mediated by conservation organizations and the dive tourism industry, who create opportunities for affective encounters unique from those that marked the past.

Multispecies encounters are mediated by the conservation voluntourism industry and through the spectacle associated with that industry. Not only does spectacle condition us to passively accept that the only path to action is through consumption, but spectacle also impedes the human ability to communicate in mutual exchanges with other bodies. That is because spectacle mimics affect; it does not enable truly affective relations, mutually constituting relations that happen *between* entities. Of course, people are affected when they consume digital media; multispecies encounters captured in digital imagery evoke emotional responses in us. But we are never able to affect the image back. This one-sided communication makes it incredibly difficult for voluntourists (and nature or dive tourists) to fully appreciate the affective encounters being packaged and presented in this new industry. In other words, the widespread consumption of digital multispecies encounters through spectacle has played such a powerful role in advancing and mediating the experiences surrounding conservation voluntourism that we do not know how to entrain—how to get in sync—with other species when presented with these opportunities.

In Ursula Le Guin's (2004, 196) famous essay on communication "Telling Is Listening," she reminds us that "successful human relationship involves entrainment—getting in sync." If we cannot get in sync, the relationship is an awful disaster, or at the very least, uncomfortable. Synchronization applies to any two things, not just human bodies. Le Guin (2004, 195) points out that clock pendulums mounted next to each other on the same wall will eventually synchronize and swing together.

> Any two things that oscillate at about the same interval, if they're physically near each other, will gradually tend to lock in and pulse at exactly the same interval. Things are lazy. It takes less energy to pulse cooperatively than to pulse in opposition . . . All living things are oscillators. We vibrate. Amoeba or human, we pulse, move rhythmically, change rhythmically; we keep time . . . this constant, delicate, complex throbbing is the process of life itself made visible. We huge many-celled creatures have to coordinate millions of different oscillation frequencies, and interactions among frequencies, in our bodies and our environment.

There are internal oscillations, such as the heartbeat. There are longer bodily rhythms of eating, sleeping, digesting, and so on. And there are even longer bodily rhythms that humans may not always be aware of, that are connected to the environment, to the seasons and moon cycle (Le Guin 2004, 195).

In "Chasing Whales with Bateson and Daniel," Katja Neves (2005) offers an account of successful human-nonhuman entrainment, describing how Azorean whalers and sperm whales navigate the waters together, interacting in ways that reveal both their similarities and their distinctions. They are connected through the ocean, through patterned movements that are "similar" but "not the same." Neves (2005, n.p.) shares the words of Daniel, a whale hunter, to illustrate her argument: "We are companions" . . . "the pattern of our movements is similar." At the same time . . . "we are both whale-like but in very different ways"; "the pattern of our movement is similar and linked through rhythmic harmony but is not the same [hence the word harmony]."

Whalers' awareness of the patterns and relations of themselves and the sperm whales, the "bodily-sensory exchange of information" that exists between them, enables the whalers to "recognize themselves in the whale" through a "process of intra-species and cross-species communication whereby whales . . . discover relationally what it is to be a whaler" (2005, n.p.). This type of affective encounter through multispecies engagement—this becoming with—is something offered up by conservation organizations and dive shops catering to multispecies voluntourists and nature tourists. The trouble is, when one-way digital transmission becomes so prevalent, as it has for the generation of tourists that dominate the conservation tourism industry, the ability to entrain, to get into the bodily rhythm with other beings, is significantly weakened.

When something is sent off into virtual space, maybe someone will see it, maybe millions of people will; or, maybe nobody will. This is because "transmission via print and the media is one-way; it's mutual or hopeful" (Le Guin 2004, 192). Maybe someone will respond; maybe nobody will. Maybe it can be the building blocks for community. But maybe not. Relationships and community can be built through these forms of community, but they are not immediate. Whatever communities that are created through digital media (printed word, filmed speech, emails, Instagram, Facebook, and so on) are exclusively mental communities. They are virtual (Le Guin 2004). Conservation and nature tourists have consumed so much of these digital sources, and been so conditioned by them, that they spend most of their time producing more imagery to contribute to the spectacle that has mediated their own multispecies encounters, rather than they do opening themselves to the emergent ecologies that mark conservation voluntourism settings, to listening to how other species are "talking" to them, to communicating, entraining, and to becoming with. Still, becoming with is one of the goals of multispecies conservation voluntourism, or at least it appears to be, even if it is not always articulated in this precise way. It is an industry that enables the generation of affect. It creates opportunities to entrain, or at a minimum, to care.

Some of the opportunities to entrain within the landscape of conservation appear more obvious than others, such as the slow observational work of species' behaviors and movements, including snorkeling to photograph whale shark patterns or monitoring and observing nesting turtles. While some multispecies voluntourism encounters are less invasive than others, I have been struck by how much of the affective labor of conservation involves capture, handling, poking, prodding, picking, cutting, and killing. And, no matter what form of care work one is involved in, it is always being documented all along the way through photographs, video, and field notes, which intercede in one's ability to entrain.

I turn now to close this chapter with some fieldwork experiences I had with organizations that presented me with opportunities to engage in affective relations and to think about life, death, and collaboration in Utila's changing landscape.

JULY 2016

We arrive at the botanical garden of one of the island's Old Heads. Antonio had been called earlier that day because Mr. Richard has five boas to be checked over, removed of ticks, and released back into the bush. Mr. Richard has been paying other locals up to two hundred lempiras to bring him any boa a hunter encounters in the bush rather than killing them or capturing to sell into the international wildlife trade market. He thinks he is seeing an increase in their population as a result. Mr. Richard now calls Antonio every time someone brings him a boa, and they film their activities for the local TV station; together they are working on a small series they call "Discovery Channel Utila," which they run after the local news. Mr. Richard holds a handheld camcorder and records our activities, introducing me as an anthropologist on the island learning about local history, culture, and nature.

We are outside a large cage that holds five boas of varying sizes. I do not like boas. The other night there was a boa wrapped around the tree less than three feet outside of my bathroom. The same bathroom that has at least four resident tarantulas (tarantulas are territorial). I don't like tarantulas either . . . I especially don't like boas and tarantulas when I don't have my contact lenses in and cannot really see their movements . . . I try and channel Mr. Richard at night when I am with limited sight, as he says these are both positive species and all houses should keep boas in their attics. They help with the rats. And tarantulas help with the cockroaches and other unwanted insects. It's a relational world that assembles together in acts of care. Still, they hide near our shoes and sometimes get in the house and my daughter worries that they will climb in our shared bed . . .

I am blown away by the two young women, one a seasonal staff person from the US and the other a conservation voluntourist from the UK pursuing

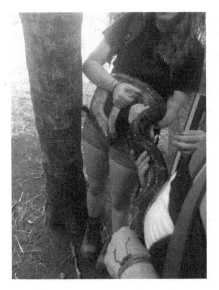
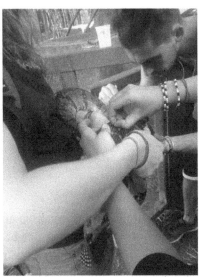

Figure 16 Extracting first boa constrictor from cage, preparing to remove ticks. Photo by author.

Figure 17 Removing ticks from boa constrictor by tweezers and fingernails. Photo by author.

her thesis research, who fearlessly enter the cage to extract the boas. The snakes are agitated by their captivity and even more so by the handling. The women calm them and speak softly to them, explaining that they mean no harm, and are there to help. We are there to remove ticks from their bodies with a small tweezer and inspect the rest of their bodies for injury (figures 16 and 17). Ticks are known vectors of hemogregarine infections and can cause rickettsial diseases in reptiles, birds, and humans.[1] Antonio offers the volunteers present that day with the opportunity to hold one of the boas. I decline the offer. They need someone to take photographs after all ... Once completed, we return the snakes to the cage for the time being. Antonio will return later in the day with some pillowcases and an ATV to drive the boas back out into the bush, placing them each in a separate location to honor the boa constrictor's preference for solitary living.

Caring for these boas by removing their ticks (they are covered with them!) not only attends to their wellbeing but also for the health of a plethora of other beings on this small island. This affective labor is invisible to most islanders, but for Mr. Richard, Antonio, and several of the organization's volunteers who meticulously and gently inspect each snake and remove the dozens of blood-sucking parasites that have attached themselves to their skin, this labor contributes to the survival of the boa, saving it from death or capture and sale on the international wildlife black market. And, this experience takes on exchange value for the voluntourists who trade the experience later on in the global market for educational scholarships or job opportunities. As well, there is a local economic alternative for hunters in the bush that is created in the capture and delivery of boas to Mr. Richard, who then offers the affective labor to volunteers.

MARCH 2017

DJ and I spend the afternoon with Diana, a young woman from the U.S. working with one of the conservation organizations to build their terrestrial programming. I had asked to observe any of the projects going on today since I was already going to be spending the morning with them, sharing back the results of a survey we codeveloped to assess the motivations and experiences of their volunteers, and brainstorming areas of improvement. Organizational staff suggested we join Diana, who was trying to launch a study of the island's invertebrate population. Within moments of our time with her, it felt suspiciously like she was crafting an experience and labor just for us. That is, I don't think she planned to go out to net today, but we were there and curious about the work, and so she did . . .

Diana provides us with large nets that we will use in time, but first she takes us to her "wet lab," where she hopes to create a cricket farm and eventually make and sell cricket-based edibles. We "toured" the lab, which was a wooden structure the size of a small shed with wood plank counters along the walls. One of the planks held two large plastic containers, which Diana said she obtained to start the cricket farm. Diana takes two bags of flour out of her backpack, opens them, and leaves them in the plastic tubs. She tells us they will be used to begin making mealworms.

We leave the lab within just a few minutes of our arrival and head to the fresh water cave off of Pumpkin Hill beach. The amount of trash on the beach is much higher than I have seen in the past. I suppose this is because we are here during the low season, low season for both tourists and nesting turtles. Beach cleanups are about to begin, focusing exclusively on these shores in preparation for turtle season. A truck will pick up volunteers once a week in town to spend the day picking up all the plastic and other refuse that has washed ashore. (No notice of nesting activity to date, by the way. Also, the former head of BICA's turtle program, is no longer there—a new person will be arriving soon to do the job. I also heard from BICA that tourists have been complaining about the monitoring activities: it is hot and full of sandflies. BICA staff said they need to think about how to enhance these experiences for the volunteers.)

We walk through the bush to a small freshwater cave that DJ and I had never before explored, enter, and dip our hands in the cool water. Diana encourages us to take a full dip in the water if we want, to allow the cool water to affect our being, to become part of the landscape and imagine those who passed through this cave before us. We don't go in the water, but we do take several photos. We probably spend about ten minutes in the cave, mostly just chatting socially with Diana, and then head up Pumpkin Hill to the same place where the orchid bee project I witnessed in 2016 took place. The hillside was completely overgrown, suggesting to me that the orchid bee study had concluded and no other island scientists were doing regular research in this particular location.

Diana demonstrates how we are to perform "sweep netting," which involves fiercely slamming our nets on the bushes and catching whatever we can, which we then shake into a small Tupperware container. We did this all along her transect, which was unmarked, but Diana noted she knew where it started and ended by sight. We walked along netting whatever we could, moving in one direction across the top of the hillside. It was but not terribly systematic: at times we would all be on top of one another, and other times, we would skip several feet of the bush, and we each collected at different heights, depending on our own height and ability. DJ and I ask questions along the way . . . We are not quite sure we understand the parameters of the study, but we maintain our enthusiasm and positive energy, engagement and curiosity, as Diana is clearly making a major effort to provide a good experience for us. She jokes and laughs a lot, while also demonstrating her knowledge of the flora and fauna up on Pumpkin Hill and of the lack of data on the biodiversity in the area (hence why we need to conduct this insect sweep and catalogue our findings).

The transect netting lasts maybe ten minutes, about as long as we spend taking photos. Once the transect is complete, we hike back down the hillside and make our way back to Diana's home base. DJ and I ask often what we are catching in our nets and what will happen to these little creatures in the Tupperware once we finish. I have a hard time tracking the types of insects Diana names. She says she will be pinning them up and sending us photos the next day (she never did, but we did see a photo of us posted on the organization's social media within just hours of our excursion, figure 18). Throughout our time Diana fretted over if she would kill them or not. She tells us that you cannot kill them if you don't have a permit (and she did not), but it is okay for people to kill for a collection . . . Once again, I see death as a central theme in conservation care work. Death is coupled with lightness, the ultimate tourist experience in Utila's affect economy.

Figure 18 Author and her husband departing for sweep netting. Photo by Ashley Adams.

AUGUST 2019

Today we took a boat out to the north side of the island to accompany Alice for more data collection [the same scientist introduced at the start of the book]. Everyone from the organization goes, which means two permanent staff members, two "seasonal staff" (volunteers), my daughter Amalie, and Leo, a student who joined me from Memphis. Leo, Amalie, and I are the first to arrive and as we wait on the organization staff, our boat captain, Edwin, a black islander, incorrectly states several times that he is waiting for a woman from BICA. [I was not with BICA that day.] I suggest we may be talking about the same woman but that she is with a different organization. He seems both unaware and uninterested in the notion that there may be multiple conservation organizations on the island.

We ride out to the Turtle Harbour refuge, passing our time chatting about the history of the island with my student, Alice, and Edwin. It was Leo's second day on the island, and he wants to learn as much as he can about its unique cultural history. Edwin tells about his experiences encountering "Indian remains" out on his trips to the north side—he says the surface continues to be littered with bones and other material remains.

Passage to the shore is difficult within the protected area, and we all must climb out of the boat in a few feet of water. Getting wet is part of the experience, and this was the first of several times we would wade through water that day. Our group splits up, with the male organization staff heading off to catch iguanas for Alice while the female seasonal staff member naps on the beach. I learn later that she isn't feeling well after a late night out, and I assume she is fighting off a hangover.

Within just a few minutes, Noel, the permanent staff member from the UK, has already caught two swampers. Amalie, Leo, and I begin to help Alice process them, and before those two are even done, three more appear, each tied up in a separate pillowcase. We spend an hour and a half collecting the data from these five swampers, handling them in the order they were caught such that they spend less time in their pillowcases. As always, Alice does a beautiful job including volunteers (us) in her work, teaching throughout the process about what we are seeing, doing, and feeling.

We measure, weigh, and sex each of them. This time Alice clips off scales rather than toes, as well as beads and inserts GPS trackers in each of them, all done to track their movements and so they can be identified if encountered again in the future. The GPS tracker is injected, and we then superglue the swamper's skin back together. It all feels very invasive, but less so than the week prior.

Last week, we were out helping Alice collect data for a colleague who is studying the highlander population. She agreed to collect 20–30 blood samples from Utila's highlander so her colleague can compare the data to samples she is collecting from the Costa Rican highlander population. The following description is from my notes that day: Alice's kit bag is very disorganized, and we have a hard time finding the items we need. Her colleague volunteered to pack the kit bag this morning but Alice didn't doublecheck it before leaving, and now she doesn't have the correct weights so we cannot weigh them. We also run out of wire so we cannot tag most of them with beads. The disorganization means we take longer to process each iguana, constantly searching for needles, beads, and clean containers to hold the toes that we clip off. Beading the first one was a bit hard to watch. The piercing with the beads went in smoothly but then when we tried to clamp the ends, Alice accidentally pulled the full string of beads through its neck, bringing blood. She then redid it and cut off two toes.

That was my first time observing Alice taking blood samples, and one of the few times Alice had done it herself. She attempts to take the first sample with a syringe. The needle goes in easily, and I write in my field jottings that the swamper doesn't seem to feel it. But how really did I know this? I was simply assuming it didn't because I did not see any physical resistance. Alice keeps the swamper's head inside the pillowcase throughout this process, to keep him calm, which probably contributed to my belief in the moment that the swamper wasn't in physical pain ... Alice keeps poking the syringe around to try to find blood, but to no avail. Eventually she gives up with the syringe and instead obtains a blood sample by pressing on the toe we just cut off—very hard, so it seems—squeezing it to make the blood flow. It feels very invasive, and this part is hard to watch. Amalie asks if it hurts them, and Alice says no—it's like cutting off a fingernail since their skin is so tough. But yet there is blood ... This happens again and again with the six highlanders. We always try the syringe first, but only in one can Alice draw blood. We learn that they can slow their blood flow when they feel threatened, so the longer they are bagged, the more likely we are not going to see blood. The pillowcase, then, serves as both a threat and a comfort? I struggle with this logic internally but place it next to the obvious care in Alice's handling of each iguana she meets, her gentle inspection and observation of their unique colors and markings, concern and commentary if one appears to have been in a fight.

Since we ran out of clean containers for the toes, we did not take any after the first two for genetic testing (figures 19 and 20). We still cut them off anyway

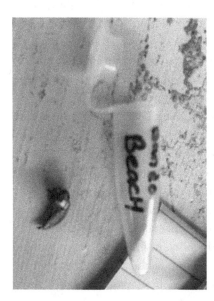
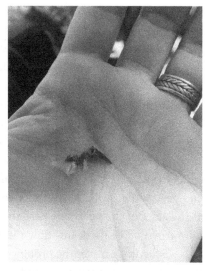

Figure 19 Highlander toe collected and saved for genetic testing. Photo by author.

Figure 20 Iguana toe to be discarded. Cut for future tracking purposes. Photo by author.

Figure 21 Author performing iguana measurements. Photo by Ivan Ortiz.

Figure 22 Logging measurements. We always take a photo of the data next to the individual iguana. The bead tag is evident in the photo. Photo by author.

and tried to extract blood that way. Then we just toss the toes on the ground. Toe removal also helps with future tracking. For instance, if you remove toes 9 and 10, the iguana number is recorded as 2019. If we cut off 7 and 10, it is 2017. One of those we found this year already had two of his toes cut off (8 and 10) and Alice was able to link him to her dataset from last year. She is thankful he survived another year in this increasingly threatened landscape.

Returning now to our time in Turtle Harbour: Leo is folded into the data collection learning how to measure and weigh the swampers. He observes me performing these tasks first while Amalie logs measurements (figures 21 and 22). We measure snout to cloacal vent for body length (SNL) and then snout to vent plus tail (STL). If the tail has broken off (usually during a fight) and begun to grow back, we take two tail measurements, one from the line at the start of the belly to where the new tail starts, and then the shorter new growth. We mark this down in the book in millimeters, for example 340 + 54. Alice handles the probing to determine their sex and the piercing through the top of their spiny necks to bead them. We volunteers take turns selecting the color sequence.

Two of the swampers we encounter are a beautiful turquoise, which Alice says indicates they are in mating season. Males turn bright blue-green hues as an attempt to attract mates. I am reminded of the beautiful blue orchid bee... All of the swampers that we catch that day are males. We learn it's much harder to catch females. Once again, my mind goes back to the orchid bees.

Apparently swampers also get very thin while in mating season: "They are so focused on having sex that they don't eat much," Alice jokes. We learn that Alice's research is starting to show that the swampers may be mating later, or they are having two seasons. Earlier research by iguana specialists suggested the swamper breeding season to be January and February, but Alice has been seeing juveniles in August and September, suggesting that perhaps the females that mate after Semana Santa are being selected for, ensuring their survival. (Semana Santa is a time when hunting for gravid iguanas increases because iguana eggs feature in a traditional Easter meal). This is one area in which swampers ensure their own survival in this landscape of change. But even if they adapt in this way, they still run up against loss of habitat, and raccoons.

After processing the five swampers, Alice suggests that we check out the area further down the coast, as she has yet to explore this part of the island herself. We must wade through waist deep water to get to the other location, something we all do happily. It's hot out, and the water is so refreshing. Not to mention trekking across muddy mangroves and snaking through small clear waterways to avoid stepping on seagrass make you feel like you are being folded into and part of the landscape—even while in your hiking boots... I had presumed we were going to the other side to find more iguanas, but we don't encounter any. Maybe we weren't looking for them... The staff trackers appear to be exploring the area but not tracking. We walk through water for what feels like another thirty or forty minutes and take more photos.

Here Alice teaches Leo about the various mangrove systems, explaining how red mangroves are closer to the coast, while black and white mangroves are further in, as they cannot stand as much salt as the red ones. She shows Leo the black, "thick and stinky" mangroves, and explains how they are really good at getting oxygen from the salt. She points out the small roots that stick up from the surface to take in air, operating like snorkels. Further inland we reach the white mangroves, and Alice points out the pendent spherical clusters of flowers of the Buttonwoods, which are all in bloom around us and smell lovely if you get close enough to them to offset the stench of the black

Figure 23 Mangrove propagules. Photo by author.

Figure 24 Journey through the water that hangs on author's office wall. Photo by author.

mangroves. We all take photos of the flowering buds. This living classroom impresses us, offering three mangrove species within such a short distance, and an opportunity for the organization's volunteers to experience and learn about their ecological significance and the life-affirming habitat they provide for us and so many other species.

Alice points out the mangrove propagules (figure 23), viviparous seeds that will soon drop from their parent tree. Mangrove reproduction is fascinating. Like many mammals, they are viviparous, which means they reproduce while still attached to the parent, thus bringing forth live young. Propagules are dropped in and dispersed by water. Once dropped in the water, the propagule must remain in water to complete its embryonic development.[2] We soon realize we are walking on top of baby mangrove trees... We take more photos... The mangroves appear so resistant to us, continuing to work to reclaim the landscape that has been ravaged by capitalism.

After a while exploring as far as we can into the refuge before hitting impenetrable red mangroves in one direction and jagged coral rock in another—reminiscent of Elsie Morgan's journey [from *And the Sea Shall Hide Them*]—we turn back to make our way back through the water. Alice ensures that we all get photos while making our way back. People love to be in water, and to get photographs in water, and in mud, a photo that will evoke a sensory experience. These are the very best to post on social media (or on one's office walls . . . [figure 24]). Leo later tells me that this was his favorite day on the island. He felt part of something bigger, beyond human. He learned along the way, was immersed in the landscape from mud to water to mangrove, forms of matter unfamiliar and presenting new and unique opportunities to engage and relate to other forms of life and being.

◆ ◆ ◆

The above photos and scenes are familiar sights, appearing in abundance on conservation websites, blogs, and social media. I had seen such images before my studies of voluntourism began, and through my research I became part of it. These images are frequently accompanied with evocative phrases about imminent threats to important and rare species, followed by heartfelt expressions of caring labor, love emojis, and lots of exclamation points. The production of tourism imaginaries is essential to Utila's affect economy. Conservation organizations that facilitate multispecies voluntourism rely on their posts being liked and shared and on encouraging others to imagine themselves in this place, to come to the island, to participate in global change science and to enjoy the "freedom" of a place with no rules (unless you are trying to fish or hunt). Conservation organizations work hard at crafting an experience for the volunteers. You can read that effort in the above examples, as BICA's staff commented on their need to improve turtle monitoring because their volunteers were complaining about the heat and sandflies, in Diana's detour with us to a freshwater cave where she thought we might enjoy a dip in the water, and in Alice's journey with us through the water and mangroves so we could have a fuller sensory experience. Within each of these spaces, the voluntourist feels that sense of liminality and freedom as their engagements with other species are acceptable and encouraged in ways that local people's engagements are actively discouraged. Conservation voluntourists are permitted within protected areas to collect data on endangered species; we are able to walk upon and wade within mangroves, cut toes, and relocate turtle nests. We voluntourists pay for these experiences. The consumption of such experiences funds the work of the conservation organizations, which in turn produces the data that the organizations need to justify further grants, to produce more science attached to the species, which then funnels back to the IUCN, informing and legitimating protected status for Utila's interesting species. Spectacle is ever present.

Through their own Instagram photos, blog posts, and Twitter and Facebook feeds, individuals are joining the contemporary moment where "green consumption is the new activism," demonstrating their engagement with fixing the problem through participating in the (market) solution (Waitt, Figueroa, and Nagle 2014, 169). This economy is fed by the selling of experiences in "becoming with" (Haraway 2008, 3), in shaping humans through their relationships of care with other, rarely encountered species. But as spectacle, other localized relationships are obscured, even to those who are managing the operations and producing the data to support continued conservation science. On more than one occasion, in the years that I have been following these organizations, I learned from Utilian-born trackers during treks through the mangroves just how "tasty" the swamper is. Here you have local guides simultaneously involved in research to preserve the species, leading expeditions for research voluntourists, and then—theoretically—going out again to track, kill, and consume one of the very creatures the organization that employs them is working to protect from human exploitation.[3] Why do it then? Even most of the scientists and conservationists I worked with on the island speculate that adding breeding programs to current conservation efforts may be a better solution to managing the swamper than a command-and-control management approach. For instance, this senior staff member of Honduran descent shared: "This is their past. It brings up memories of their childhood. How can I tell someone not to partake in their cultural heritage? I think we should be breeding the swamper so islanders can eat it, not merely for scientific observation and preservation."

Why do it then? Because organizations have figured out that they can be sustained financially if they transform their roles into "affect generators" and create opportunities for the affective exchanges people are after in the twenty-first-century economy. The interest of people to engage in relations with other beings through emotional and affective labor can be explained by changes in the forms of labor valued in the twenty-first-century global economy, and Alan Bryman's (2009) concept of "Disneyization" is instructive in explaining this transition. Bryman argues that the growing desires to engage in emotional and affective labor emerge from transformations in society that have stripped away and hidden the original or real nature of a place or event, presenting a sanitized, pleasant, and easy-to-grasp experience: a Disneyified experience.

Bryman describes four trends that carry over from Disney theme park motif into the larger world, which then influence the ways in which our lives take shape and the content of our existence and engagements. The first is "theming." This is where previously disparate elements combine into one coherent, *fun* image. Examples of theming include resorts, shopping centers, and even now university campuses and national parks and protected areas. Second is the "dedifferentiation of consumption." Consumption that had formally taken place in distinct spheres now is joined together, for instance, with theme parks selling merchandise alongside food, rides, and entertainment. Associated with the convergence of spheres is the third theme, "merchandising," with goods now manufactured for the sole purpose of sales (think of Disney character collectibles, national park gift shop items, or "save the swamper" T-shirts). Finally, the fourth trend is the rise of "emotional labor." This service work is scripted to include cheerful friendliness; this feigned emotional labor has the effect of distracting customers from the fact that they are being captivated into an artificial environment. As Bryman (2009, 66) writes, "The ever-smiling Disney theme park employee has become the stereotype of modern culture. Their demeanor coupled with the distinctive Disney language is designed among other things to convey the impression that employees are having fun too and therefore not engaging in real work."

This is the type of labor that marks multispecies conservation voluntourism, and it is the form of labor valued in the new economy. The demeanor of the happy conservationist–turned–tour guide appears throughout social media sites of these organizations, which then invite future voluntourists to join in efforts to protect and save rare or endangered species from other species, including human residents of far-off destinations. These images serve as invitations to general tourists who are attracted to imaginations of discovery and the sublime, with desires to experience transcendence underwater with other beings.

The desire for a less alienated relationship with the natural world is coupled with an appeal of being able to exchange these experiences for status. Money purchases the opportunities to engage in multispecies encounters, and these experiences are then exchanged through social media posts and resume lines for something else, usually occupations with higher income generation. The payoff of conservation voluntourism in Utila is not too shabby. For instance, a former staff member from WSORC who came through their internship program is now being supported for doctoral studies at an Ivy institution in the states.

The opportunities to engage in affective labor with other species, to potentially become with, that are created in this emerging green economy take on exchange-value, keeping Utila integrated into the global market. Yet, the emphasis voluntourists place on getting the right photos to document their engagement, to extracting and recording data from species (toenail clippings, photographs of whale shark spots), suggests that the privileged position of the conservation volunteer to experience and become with might be lost due to their conditioning by spectacle. The focus on data, documentation, and dissemination to social media networks may get in the way of being able to fully entrain, to get into sync and to become with, at least in the way conceptualized in the emergent ecologies literature. It may limit a voluntourist's ability to think beyond the one-world world and to appreciate alternative worldings in Utila and beyond. But perhaps not. Perhaps the opportunity to immerse and entangle oneself with other life beings allows us to imagine life beyond the Anthropocene, a life where human exceptionality and domination is no longer taken for granted, a proposition taken up in the next and final chapter.

Conclusion

Life, Death, and Collaboration in Utila's Affect Economy

I wish to return to one of the questions that framed this book, as I sat on Pumpkin Hill in 2016 with the orchid bee team, thinking about Donna Haraway's (2008) reminder about how foolish we humans are to think we are in any way exceptional and to instead embrace the future through an understanding that "becoming is always becoming *with*, in a contact zone where the outcome, where who is in the world, is at stake." I wondered back then on that hill, who and what will persist in this landscape a century from now? I still wonder this, and while I obviously don't have the answer—and I barely understand the biological and ecological aspects of the other beings I have encountered in Utila—I do have some scholars I look to for inspiration to think through possibilities for an abundant future that reckons with its colonial past.

Embracing the "affective turn" in the social sciences, decolonial feminist scholars are offering new ways of thinking through the urgencies of the Anthropocene (e.g., Barad 2007; Haraway 2016; Tsing 2015). To begin this journey, we might consider Anna Tsing's (2015) journey in *The Mushroom at the End of the World: On the Possibility of Life in Capitalist Ruins*. She offers the mushroom as an example of what emerges in ruined landscapes, places ravaged by capitalism. The mushroom feeds off and thrives on dead organic matter; it is an example of the "contaminated diversity" that can emerge in twenty-first-century landscapes. Her story of those who harvest and sell the matsutake is an example of "contamination as collaboration," a story of collaborative survival within multispecies landscapes devastated by capitalist destruction.

Similarly, Donna Haraway's (2016) *Staying with Trouble* offers us a new perspective in conceptualizing our relations to the earth and all of its beings. She upends the Anthropocene, replacing it with a new label for our times, "Chthulucene," which enables a conceptual shift from human exceptionalism and utilitarian individualism to making and becoming with. "Chthulucene" is a compound word of two Greek roots that together describe a time of now, beginnings, and freshness plus unruly beings of the earth. Conceptually, it means approaching the now as multispecies kin relations, as ongoing presences with relationships and life of "all sorts of temporalities and materialities" (Haraway 2016, 2). Haraway (2016, 2) argues that "living with and dying-with each other potentially in the Chthulucene can be a fierce reply to the dictates of both Anthropos and Capital." Utila's environs are another story of collaborative survival, of many different species, in landscapes ravaged by capitalism, and this book a critique of the Anthropocene concept.

We see collaboration through possibilities created through the lionfish, which is out of place and also now increasingly in place. For instance, the lionfish-culling program may not be producing the conservation goal the organizations involved hope to in terms of the diversity of the marine ecosystem, but it is producing value through the affective labor of volunteers with Utilian hosts. The lionfish-culling program creates data for conservation volunteers who are studying the lionfish's feeding preferences through gut content analysis. It also contributes to locally hosted lionfish derbies, where island restaurants engage in a cook-off competition and party attended by tourists at least twice a year, sometimes more. The lionfish culling and derbies create opportunities for affective exchanges between hosts and guests, both human and nonhuman within the marine environment, as humans care for the survival of other species through the death of the lionfish. Thus, the lionfish is valued in Utila's newly emerging economy through its death and in part as a tourist attraction. Within Utila's emerging multispecies voluntourist economy, there is value now in the death of lionfish and in the survival of the iguana, or the turtle.

As we attend to the agency and autonomy of other beings that are cohabiting and collaborating for survival in a landscape wrecked by capitalist destruction, we begin to ask about raccoons and how their efforts to survive may alter the future for other species, including blue crab and iguana. And what will the relationship between the raccoon, iguana, and crab mean for humans on the island who have relied on these food sources as part of their sustenance? And as we think about the agency of nonhuman beings, we are reminded of the resilience of propagules out in Turtle Harbor and beyond. Will these mangrove babies be able to push through and serve as swamper and blue crab breeding grounds into the future? And, if so, how soon will they then be cleared, infilled, and transformed again as spaces of possibility for new island residents seeking refuge and opportunity from mainland violence or as fantasy homes for lifestyle migrants?

There are other questions that arise when we consider the future of Utila. If we embrace the new materialist call to move beyond a narrow and anthropocentric understanding of the human condition, one that returns to Maturana and Verden-Zöller's (2008) "biology of love," and instead see the world as unstable assemblages that are constantly in a state of emerging, we must confront the patriarchal history of colonization and dispossession of both humans and nonhuman species. In so doing, we realize that the brief history described in these pages is just that: brief. It is a story born out of colonialism, a settler narrative written from a primarily Eurocentric, masculine frame.

We know very little about the multispecies assemblages beyond those existing at the time the Pech were on the island and Cays beyond those appearing in colonial writings. We know a little about the settler population's multispecies interactions at the turn of the century thanks to the writings and oral histories presented in chapter 2. We know sharks and alligators were once plentiful but have all but disappeared. We know that the fisheries were far more abundant, turtles were a way of life, and Old Tom was a fixture from the very start of settlement in the lives of the island's Old Heads. We know settler families exhibited multiple worldings, including—*but not limited to*—an ontology closely associated with modernity and its normative divides that separate out humans from other life-forms and insist on "productive use" of nature. We know by reading behind the lines of historical accounts, and by listening closely to the stories Utilians tell, that settlers and their descendants were (and are) engaged in relational and reciprocal exchanges with the lifeworld. We know there are multiple forms of worlding on Utila and beyond. By following a political ontology approach, we also know that there is power, conflict, and struggle in world making. But there is also creativity and imagination. We must, therefore, attend to the history of colonialism, accumulation, and dispossession, just as we attend to futures and possibilities.

To reckon with the colonial and settler history of this small island space is to recognize that humans may be constantly attempting to shape and narrate multispecies assemblages, but in actuality humans are not all that exceptional. Nor will humans realistically be present in the same form in the near future of the earth. To this end, Deumert's (2019) call to "think-with-the-mangrove" embraces Haraway's conceptional shift from the Anthropocene to the Chthulucene in a way that we humans avoid the narrow lens of utilitarian individualism and embrace the agency of the nonhuman. To think-with-the-mangrove allows us to see the new entanglements emerging within Utila's affect economy, ones that offer fresh beginnings and collaborations. Sometimes new possibilities emerge through the possibility and experience of death and destruction.

Right now, we are seeing new human actors making decisions about life, death, and collaboration on the island. Tourist desires and conservation science facilitate and encourage the death of some species for the future of others (e.g., lionfish death for reef fish survival), or even the death of some of one's own to ensure the future survival of their offspring (as in the case of the orchid bee). Nonhuman species also have a role in determining the future of life and death on the island: in this short history of just slightly more than 100 years since British settlement, the coconut gardens were continuously destroyed by insects, from the "coconut bug" of the early 1900s to the lethal yellowing disease of the early 2000s that was spread by plant-hopping insects. Now the lionfish are transforming the reef, and raccoons are eliminating the crab and iguana. To think about who and what may live and die in Utila's affect economy, we must contend with the agency of nonhuman species and how their movements are influencing life, death, and collaboration.

Notes

Introduction

1. "Archipelago" technically refers to the sea that surrounds islands but has come to be used today to refer to islands. The word is derived from Italian *archi* (chief or principal) + Greek *pelagus* (pool or sea).
2. Others working in this field have referred to this industry as ecological voluntourism (e.g., Waitt, Figueroa, and Nagle 2014).
3. The researchers used "invert team" as shorthand for the group focused on research with invertebrates (animals lacking a backbone).
4. Pseudonyms are used throughout the book except in cases of public figures.
5. As with the "invert team," the "vert team" was a shorthand way to refer to individuals whose research focused on vertebrates, or animals with backbones (e.g., birds, reptiles, and mammals).
6. While not a focus of this book, elsewhere I am working through the connections that exist between gender, custodial labor, and the production of affect. Parreñas (2012, 674) defines custodial labor as "the process in which affective encounters between bodies fill a demand for meaningful purpose among professional workers (usually from the Global North) who engage in commercial volunteerism or other efforts that at first glance appear to be altruistic." Multispecies conservation voluntourism enables women to engage in affective relations by performing "custodial labor" alongside other gendered bodies. But, as we will see in this book, other gender dynamics are also at play, including masculine acts of discovery, entitlement, and exploitation of landscapes and bodies.
7. Researchers on the island tag swampers in three ways to track their movements or in case they or another researcher encounters the same iguana again in the future. The least expensive method is by painting an identification number with

nail polish. A slightly more expensive and invasive method is by inserting beaded safety pins with uniquely coloured beads into their spines for identification purposes. The final and most expensive way is to inject a chip for GPS tracking.
8. "Mestizo" is used to refer to individuals of mixed Spanish and Indigenous descent. The identifier "black" is used in the text to refer to Afro-Caribbean descendants and "brown" to individuals of mixed descent. For more information, see Wade (2010) and Rahier (2012).
9. See Harcourt and Nelson (2015); Mostafanezhad et al. (2017); Peet, Robbins, and Watts (2011); Robbins (2019); Rocheleau, Thomas-Slayter, and Wangari (2013); Stonich (1999); and Watts and Peet (2004).
10. E.g., Brondo (2013); Douglas (2014); Hoffman (2014); Stonich (1998, 1999); and West, Igoe, and Brockington (2006).
11. See, e.g., Barad (2007); Chrulew et al. (2012); Haraway (2008, 2016); Kirksey (2012, 2014, 2015); Kohn (2007, 2013); Johnson (2018); Schulz (2017); Tsing (2012, 2015).
12. Emerging in the mid-2000s, the critical literature theorizing the relationship between neoliberal market approaches and the proliferation of protected areas for biodiversity conservation is now quite vast and includes many studies that explore the negative consequences of commercialized conservation. E.g., see Arsel and Büscher (2012); Berlanga and Faust (2007); Brockington and Duffy (2011); Brockington, Duffy, and Igoe (2009); Brondo (2013); Brondo and Bown (2011); Büscher and Dressler (2007); Büscher, Dressler, and Fletcher (2014); Devine (2014); Fletcher (2010); Grandia (2007); Holmes and Cavanagh (2016); Igoe and Brockington (2007); Sullivan (2013); and Sullivan, Igoe, and Büscher (2013).
13. See, e.g., Arsel and Büscher (2012); Büscher, Dressler, and Fletcher (2014); Büscher et al. (2012); Igoe and Sullivan (2008); and Roth and Dressler (2012).
14. Marx ([1867] 1957) distinguished use value, which refers to the usefulness of a commodity, from exchange value, which is the value of a good or service in the open market. A commodity's "use" is tied to its physical properties, or the ways in which an object can be put to use to fulfill human needs. Exchange value is determined in comparison with other objects on the market. In capitalist markets, exchange value dominates, as items that are exchanged in the market are always compared to a third category—money—which is taken to be the universal equivalent of the item. Thus, exchange value is the value that an item or service holds in economic terms. This object may not actually be useful at all (in terms of satisfying a human need); it may simply hold value on the market due to trends or rarity. With the emergence of capitalism, we find that what the "labor theory of value" no longer defines the price of a good or service. That is, value is no longer defined by "socially necessary labor" or the total amount of labor time required by society to produce a good or service. We carry this production of value forward today, and it is crucial to our understanding of the ways in which the "affect economy" that I am describing operates.

15. E.g., Caissie and Halpenny (2003); Lyons and Wearing (2008); McGehee (2002); McGehee and Norman (2002); McGehee and Santos (2005); Ooi and Laing (2009); Tomazos and Butler (2009); Wearing (2001, 2002); and Wearing, Deville, and Lyons (2008).
16. The forces that have led to the blossoming of this sector also emerge out the effects of a global adoption of neoliberal ideology and the global restructuring of people-nature relationships under neoliberalism. NGOs have arisen to "fill in the gaps" (Fisher 1997; Schuller 2009) left by the withdrawal of the state from environmental management. These organizations are reliant on external funding to keep themselves afloat, and the growing voluntourism sector offers another revenue source to the unpredictability of granting programs. This sector continues to grow, as global citizens work to accumulate the cultural capital and competencies that are valued under twenty-first-century neoliberalism.
17. Comparing volunteer opportunities of the past (such as the Peace Corps) to those of today, Butcher (2011, 75) writes that "volunteer tourism today is a personal and lifestyle strategy to make a difference in the world," finding volunteers of the twenty-first century to possess an "individualistic, narcissistic, and incredibly limited approach to politics." Although the experiences captured in this literature review are individualized and impacts seem largely centered on the "guests" over the "hosts," there still may be a benefit to "developing people," who throughout the course of their lives and careers may become more attentive to acting in ways that will benefit those less well off than themselves.
18. "Doing it for the gram" is slang for when an individual does something for the sole purpose of posting a photo or video of the act to Instagram.
19. The 2011 summer fieldwork was focused on volunteers and the work of conservation organizations and the municipality's environmental agency. During this phase of the project I conducted in-depth semistructured interviews with local conservationists (NGO staff and officials from the municipality) and dozens of volunteers at the Iguana Research and Breeding Station. I also worked as a volunteer with the Iguana Station, participating in a range of activities including gathering and preparing food to feed captive iguanas (i.e., fruits, hibiscus flowers, termite nests, and crabs), cleaning cages, leading tours for Iguana Station visitors, searching mangrove forests to capture and release swampers (which were weighed, measured, and tagged for further tracking), and interacting with schoolchildren from the island during Iguana Station–supported activities. Beyond the research with conservation professionals and volunteers, as in other years, I also engaged locally, alongside my partner and two children, with community members at local events, and simply "hanging around," visiting with people when the ferry or airplanes arrived from the mainland, at the public beach, and among households or in local business establishments.
20. Surveys included a survey of dive shop owners to assess their willingness to participate in promoting a voluntary visitor fee for divers and snorkelers to enter the protected waters surrounding Utila, and several unique surveys of conser-

vation volunteers exploring topics including the relationship between gender, motivations for volunteering, and environmental values (totaling 168 completed surveys of conservation volunteers), as well as more evaluative surveys to help the organizations improve their programming.
21. I also led a research team responsible for fifty-three individual interviews with conservation staff, local government, and key informants from the local community, and eighty door-to-door sustainable livelihood surveys.
22. On June 28, 2009, the Honduran military orchestrated a coup d'état to depose President Manuel Zelaya. Before the 2009 coup, Honduras already had a bad track record for human rights abuses. After that time, the country escalated and, according to the UN, became "the most dangerous (peace time) country" on the planet by 2011. Despite a declining trend in violent crimes, the murder rate in Honduras still remains among the highest in the world. Much of the violence is concentrated on the mainland.

Chapter 1

1. The origin of the name was told to me by a local woman whose childhood predated the neighborhood. I have never seen this documented elsewhere.
2. There were two turtle projects, but once Honduras signed on to the Inter-American Convention (IAC) for the Protection and Conservation of Sea Turtles, BICA agreed to discontinue these projects. As endangered species, the international conservation movement did not want local people without the proper credentials keeping turtles in captivity, even if, as one local shared, they had just as much knowledge—perhaps, arguably more—than trained conservation professionals.
3. Plastic consumption contributes to a startling number of turtle deaths each year. The Sea Turtle Conservancy estimates that over 1 million marine animals (mammals, birds, sharks, fish, and turtles) die each year as a result of plastic consumption. "Information about Sea Turtles" (n.d.).
4. Marine biologists working in the region speculate that the continuing loss of beach habitat due to building and erosion occurring on Pumpkin Hill Beach will soon impact not just nesting activity in Utila but also that within the entirety of the Bay Islands Marine Park. In 2018 and 2019, BICA Utila reported seeing turtles on Pumpkin Hill Beach that had been tagged by scientists from ProTECTOR who were monitoring the nesting turtles at Sandy Bay West End Marine Reserve off of Roatán. These sightings suggest that turtles from the juvenile population in Roatán are maturing and possibly nesting on Utila's beaches. Marine biologists from ProTECTOR speculate that it may then be possible that hatchlings from Pumpkin Hill Beach are recruiting into the marine reserve off Roatán and are part of the supply of juveniles into the reserve. If this is the case—if nesting on the beach reduces or stops altogether due to loss of habitat—the flow of juveniles coming into the marine reserve may slow or stop completely, all due to the link to housing and tourism development on Pumpkin Hill Beach

in Utila (Steve Dunbar, personal communication, January 6, 2020). See http://www.turtleprotector.org for information on the work of ProTECTOR; the website features downloaded research reports and reference material for journal publications based on the organization's work.
5. Local nest guardians mark the sites so they can monitor activity, and to ease the baby turtle's journey from nest to sea. The sand is smoothed out within the "landing strip" because it is difficult for baby turtles to climb over and out of the gulfs that footsteps create in the sand.
6. A handful of people have offered to take me to the cemetery, and I've remained reticent given the ethical conundrum of an anthropologist bearing witness to looting and spoliation of archaeological sites.
7. The section of the pamphlet reads: "Friday told Crusoe that he had been in a party on the north side of Utila that devoured 23 captives, 20 men, 2 women and a child. He also told Crusoe that 17 bearded whites had come ashore on the mainland of Honduras. Crusoe asked Friday if they ate them and Friday was puzzled. He said that they only ate captives after a fight" (McNab n.d., 9).
8. Mechanisms of erasure include the creation of maps and land reforms that impact distribution and ownership, resource exploration and exploitation, changes to diet, landscape use, and domination of languages. See Perley (forthcoming).
9. See Melissa Johnson's (2018) *Becoming Creole* for an analysis of how defining human in this narrow fashion (which captures only a very small percentage of the global population) dispossesses and denies other genres of human being. Johnson's work on *Becoming Creole* employs Wynter's call to move beyond the overbearing colonial and Eurocentric conception of human.
10. Lord (1975) made this observation during his 1970s fieldwork.
11. Jackson (2003) uses capital letters to show the emphasis that Utilians place at the end of their words when speaking Creole.
12. See Smedley (1993) for a historical discussion of the invention of race and whiteness. Smedley meticulously details how race and its ideology about human differences emerged in the context of African slavery.
13. In the novel, the wildness of the mangrove and bush is contrasted with the order and safety of the coconut garden and Eastern Harbor, colonized spaces that serve as a refuge from the dangers awaiting a white girl in the bush.

Chapter 2

1. Sometimes I would outright ask people during an interview how they would describe a Utilian, as this question would energize most local people, who had much pride over their history and shared culture. Sometimes I would ask the question, and sometimes people would just turn the conversation in this direction, in interviews and in normal conversation. This was an instance of the former.
2. It is as if Johnson were writing about Utilians, and as if the quote from the machine shop owner had come from her study of rural Creole Belizeans. The idea of freedom as found in the Belizean villages of Johnson's study or on Utila

and its Cays is drawn in sharp contrast to life in cities of Belize or mainland Honduras, or in the United States.

3. And in both cases, the notion of freedom as found in the bush is closely attached to the significance they place on land. Johnson (2018, 64–65) details how in Belize and other parts of the Caribbean, emancipated slaves and free black people faced restrictions on owning property as property laws were defined intentionally to keep a ruling class of landowners with a large pool of poor laborers. Thus, to secure land was a both a material and symbolic evidence of freedom from the ruling class. This history, Johnson argues, created a very close attachment to place.

4. Recent examples of local upset in Utila include anger with the municipality for installing thick blue ropes across the main concrete streets to serve as makeshift speed bumps and slow the pace of motorized traffic and for instituting a ban on plastic bags in island stores.

5. I am focusing here on the section of West's work on surfers, described in ch. 1 of *Dispossession and the Environment* (2016).

6. Based on my experiences on Utila and on the more commercialized tourist destination island of Roatán, I suspect this footage came from film footage during another HGTV trip to Roatán, where there are plenty of spider monkeys to crawl on tourist shoulders.

7. This is Carnival as experienced among tourists. Locals still enjoy Carnival celebrations and are responsible for organizing the weeklong activities and parade.

8. Dupuis, who born in Mexico City, settled in Honduras in 1993 and since that time created and published several successful English language travel magazines, including *Copan Tips* and *Honduras Tips*.

9. From content alone, I would deduce that the vast majority of comments were from tourists, past or future.

10. True, there are some jobs for locals within the tourism industry, but like most small islands reliant on small-scale dive tourism, it is extremely difficult for most to make a decent living. I return to livelihoods in chapter 4.

11. They were "fortunate" in that they were carrying sticks to scare off roaming dogs and were able to fight back and escape the assaulters; however, both were very badly injured, bloodied and bruised.

12. The consulate did come to her, and not only did they tell her she was within her right but they also said they wished they could get someone like her over in the states to enforce the law.

13. The frame presented in this section closely resembles many other narratives of change, as nostalgia runs deep in most communities. As well, it would be wrong to suggest that all Utilians are critical of the island's party scene. Indeed, many do participate, especially younger generations, and many benefit financially from the island's party reputation.

14. Spelled *gifiti*, this drink is a traditional medicinal alcoholic beverage made by soaking roots and herbs in rum of the Afro-Indigenous Garifuna.

15. The law also forbade any taxes, fees, or other form of financial obligation for cultural presentations and shows, tying the growth of tourism to the presumed social and cultural development of the Honduran people. In fact, any project that was dedicated to the preservation or restoration of cultural heritage was granted the added benefit of an exemption from municipal property tax (Decree no. 314–98).
16. While islanders held properties on the bay, they typically kept their homes higher up on the hillside for the breeze and to escape the incessant biting of sandflies.
17. Key accords to promote tourism include the 1991 Acuerdo Ministerial Número Dos, which provided minimum standards for development, and Decree 83–93 of 1993, which created a Bay Islands Commission to promote development (Stonich 1999, 67–68).
18. As one of the closest port cities in the states, many generations travel back and forth to New Orleans, or "Little Utila."
19. These processes of gendered land loss and the ways in which local politicians may have abused their positions of power and authority to ease land sales in order to encourage foreign investment closely resemble that which I documented regarding Garifuna land loss on the north coast of Honduras in the early 2000s (Brondo 2007, 2010, 2011, 2013, 2018).

Chapter 3

This film was produced independently by two underwater videographers from Germany and England; the latter worked as a scuba instructor on the island for some time. See http://www.whalesharkfilm.com/eng/index_eng.html for more details.

1. Moore is not alone in adopting the GCS label. Several interdisciplinary academic programs and international research and think tanks identify as GCS teams. See, e.g., MIT's Center for Global Change Science (https://cgcs.mit.edu/) or the U.S. government's Global Change Research Program: https://www.globalchange.gov/what-we-do/advance-global-change-science.
2. This dearth may be linked to anthropology's reticence to engage in tourism studies in general. The long-term reluctance from anthropologists to take tourism seriously as a field of study may have been because tourism was seen as largely about relaxation and play, rather than "work" (Gmelch 2010, 6). Another central reason could be that anthropologists did not want to acknowledge how similar they were to tourists themselves (Gmelch 2010). Anthropologists have long considered fieldwork a rite of passage that relied in part on the discovery of a place and people. Many anthropologists may have been hard pressed to accept that the people they studied were also visited by other, non-serious outsiders called "tourists." Malcolm Crick (1995) argued that the anthropologist's identity comes into question through tourism: since tourists are not taken seriously, then the entirety of the anthropological discipline is called into question—is the idea that anthropologists spend time "being there" enough?

But this is the anthropology of the past, a time when anthropologists also analyzed tourism in terms of "hosts" and "guests," without reckoning with the fact that "locals" and "visitors" themselves are not homogenous groups (Salazar and Graburn 2014, 15). Locals include property owners and the property-less; people of varied educational, occupational, and demographic backgrounds; people with distinct interests and stakes. Visitors include tourists, as well as investors, businesspeople, travel industry staff, and researchers. All of these actors assemble together to produce new meanings and experiences.

3. In 2020 my conservation partners, Suzanne Kent, and I began a project to attempt to invert this practice, by collaborating with local people to feature their local knowledge in school environmental education programming, supplementing programming that had been developed and delivered by environmental scientists from the mainland and outside of the country. The project launched with National Geographic Society funding as this manuscript was completed.
4. For discussion and examples of commodifying and fetishizing nature see Carrier and Macleod (2005); Igoe, Neves, and Brockington (2010); Marx ([1867] 1957); and Neves (2010).
5. Many NGOs that support these activities actively push best practice codes of conduct to try to mitigate these issues. Moreover, those who support ecotourism of this form do so in place of direct harvesting of species. Thus, they rationalize that even with the stress to the species caused from tourist encounters, the species would likely have otherwise been killed, so it is an overall biodiversity conservation gain.
6. See the following for more discussion: Brockington (2009, 2014); Igoe (2016); Igoe, Neves, and Brockington (2010, 487); Mostafanezhad (2013).
7. E.g., Fassin (2011); Feldman and Ticktin (2010); Freidus (2010); Malkki (1996, 2010); Mostafanezhad (2013); Ticktin (2011).
8. One friend refers to Mr. Shelby as a "legend" of the island, remarking that he is the nicest man to ever grace the community.
9. New arrivals still haul building materials in sacks to sell to island developers, but no longer do they dig out the beachfronts. Nowadays they head to the bush. With so much swampland to be filled in, there is a never-ending demand for rock. Primary locations for extraction are around Pumpkin Hill and the airport where ample deposits of coral fragments, limestone, and metamorphic rocks can be found. Breaking up the rock formations in the bush is hard labor but comes with a handsome payoff for those without other sources of income.
10. Blackish Point is located on the northern end of the island, within the limits of the Turtle Harbour Marine Reserve.
11. The Reef Leader program trains a select group of local young people in conservation practices and environmental education. They participate in a variety of ecological stewardship projects and are being trained to work as eco–tour guides.

12. Still, Mr. Shelby donated the land on which the current BICA office is now located, a $40,000 value, and continues to fund many conservation projects on the island.
13. The same happened in Utila's schools, where the teachers have been almost completely replaced by mainland teachers.
14. A minimum stay is required because training volunteers is time consuming. It takes at least a week to get comfortable enough in knowing the basic activities needed to keep the station operating, and much longer if you wish to become involved in the ecological research being performed.
15. Female raccoons give birth to between 1 and 8 cubs after a gestation period of between 2 and 3 months, thus potentially giving birth to 48 cubs a year.

    In 2021, conservation organizations ran a household survey to gather local feedback to inform a raccoon management plan. During this process, they learned of multiple households who brought raccoons over to the island as pets, which they later released to the bush.
16. FIB, BICA, and UMA are currently working to develop a management plan to address this problem, whether it be through capture and release on the mainland or outright culling.
17. By 2019, staff declined again. By then the Station was left with just the director, one part-time tracker, and a neighbor who has volunteered their time ever since they helped establish the German station.
18. Since the 2011 fieldwork, I have collaborated with island conservation organizations on three distinct web-based surveys of past conservation volunteers. Each unique survey confirmed similar motivations for volunteering, with the overwhelming majority of respondents always reporting that they volunteered on Utila as a means to gain experience in conservation and environmental work in order to advance their careers in a related field.
19. To be fair, I attended their volunteer training in 2016 and found it quite robust.
20. The title for this section is taken from the title of an article on GoNOMAD travel website by Driver (2020).
21. Opwall volunteers are not allowed to spear lionfish (only Opwall staff scientists can); dive tourists and conservation tourists who are trained through WSORC and BICA programs can participate in spearing. Opwall does not allow its volunteers to spear lionfish for several reasons, including the risk of them accidentally damaging the reef and because the UK university ethical review committees that they work with require killing procedures of a higher standard than the BICA guidelines; they must ensure a quick death for the lionfish, and often spearing does not result in immediate death, potentially leaving lionfish alive on the spear hours later (Andradi-Brown, personal communication, November 9, 2017).
22. The Reef Fee eventually moved from under BICA's management to the municipality. The current Reef Fee is $3 per day of diving.

23. The GoFundMe campaign, "Utila Lobster Diver Program" was created April 28, 2018. Updates to the campaign and its associated Facebook page logo were titled "Saving Lobster Divers."
24. It is also likely that a bulk of it went to film production mentioned in the GoFundMe page.
25. See Anderson (2012); Jackson and Palmer (2014); Parreñas (2012); Ruddick (2010); Singh (2013).

Chapter 4

1. See Burridge (2005) and Novakova et al. (2015) for discussion of the relationship between rickettsial infections in ticks from reptiles in Honduras and measures for controlling and eradicating tick infestations.
2. Red mangroves drift for the longest, around 40 days. Black mangrove propagules drift for 14 days and white mangroves for just 5 days, which includes germination.
3. Similar contradictions were found among locally owned businesses that hosted large groups of conservation voluntourists in that they eliminated specific, less environmentally sustainable dishes from their menu only when the conservation voluntourists were on the island.

References

AboutUtila. 2021. "Tropical Island of Utila, Bay Islands, Honduras." AboutUtila.com. http://aboututila.com.

Anderson, Ben. 2012. "Affect and Biopower: Towards a Politics of Life." *Transactions of the Institute of British Geographers* 37(1): 28–43.

Andradi-Brown, D. A., M. J. A. Vermeij, M. Slattery, M. Lesser, I. Bejarano, R. Appeldoorn, G. Goodbody-Gringley, A. D. Chequer, J. M. Pitt, C. Eddy, S. R. Smith, E. Brokovich, H. T. Pinheiro, E. Jessup, B. Shepherd, L. A. Rocha, J. Curtis-Quick, G. Eyal, T. J. Noyes, A.D. Rogers, and D. A. Exton. 2017. "Large Scale Invasion of Western Atlantic Mesophotic Reefs by Lionfish Potentially Undermines Culling-Based Management." *Biological Invasions* 19 (3): 939–54.

Anzaldúa, Gloria. 2002. "Now Let Us Shift . . . the Path of Conocimiento . . . Inner Work, Public Acts." In *This Bridge We Call Home: Radical Visions for Social Transformation*, edited by Gloria Anzaldúa and AnaLouise Keating, 540–78. New York: Routledge.

Arsel, Murat, and Bram Büscher. 2012. "Nature™ Inc.: Changes and Continuities in Neoliberal Conservation and Market-Based Environmental Policy." *Development and Change* 43 (1): 53–78. https://doi.org/10.1111/j.1467-7660.2012.01752.x.

Baackes, Alexandra. 2011. "Surviving Sunjam." *Alex in Wanderland* (blog). July 31, 2011. https://www.alexinwanderland.com/2011/07/31/surviving-sunjam/.

Barad, Karen 2007. *Meeting the Universe Halfway: Quantum Physics and the Entanglement of Matter and Meaning*. Durham, N.C.: Duke University Press.

Bay Islands Conservation Association (BICA). 2019. "Presentación del escenario actual y elaboración de la visión de desarrollo." PowerPoint presentation.

Bell, Claudia, and John Lyall. 2002. *The Accelerated Sublime: Landscape, Tourism, and Identity*. Westport, Conn.: Praeger.

Benson, Michaela, and Ka O'Reilly. 2009. "Migration and the Search for a Better Way of Life: A Critical Exploration of Lifestyle Migration." *Sociological Review* 57 (4): 608–25.

Berlanga, Maruo, and Betty B. Faust. 2007. "We Thought We Wanted a Reserve: One Community's Disillusionment with Government Conservation Management." *Conservation and Society* 5 (4): 45–477.

Blaser, Mario. 2009. "The Political Ontology of a Sustainable Hunting Program." *American Anthropologist* 11 (1): 10–20.

Blaser, Mario. 2013. "Ontological Conflicts and the Stories of Peoples in Spite of Europe: Towards a Conservation on Political Ontology." *Current Anthropology* 54 (5): 547–68.

Blog. 2016. Last accessed February 6. http://missstep.org/utila-bay-islands-honduras/.

Bolles, Lynn, A. 1992. "Sand, Sea, and the Forbidden." *Transforming Anthropology* 3 (1): 30–34.

Bolles, Lynn. 2020. "Reciprocal Arrangements: The life Story of 'Rosalind'—Woman, Worker, Mother, and Jamaican." *Feminist Anthropology* 1(2): 260–71.

Braun, Bruce. 2004. "Querying Posthumanisms." *Geoforum* 35(3): 269–73.

Braun, Bruce. 2006. "Environmental Issues: Writing a More-than-Human Urban Geography." *Progress in Human Geography* 29 (5): 635–50. https://doi.org/10.1191/0309132505ph574pr.

Brockington, Dan. 2009. *Celebrity and the Environment: Fame, Wealth, and Power in Conservation*. London: Zed.

Brockington, Dan. 2014. "The Production and Construction of Celebrity Advocacy in International Development." *Third World Quarterly* 35 (1): 88–108.

Brockington, Dan, and Rosaleen Duffy, eds. 2011. *Capitalism and Conservation*. West Sussex, UK: Wiley-Blackwell.

Brockington, Dan, Rosaleen Duffy, and Jim Igoe. 2009. *Nature Unbound: Conservation, Capitalism, and the Future of Protected Areas*. London: Routledge.

Brondo, Keri V. 2007. "Garifuna Women's Land Loss and Activism in Honduras." *Journal of International Women's Studies* 9 (1): 99–116.

Brondo, Keri V. 2010. "When Mestizo Becomes (Like) Indio . . . Or Is It Garifuna?: Negotiating Indigeneity and 'Making Place' on Honduras' North Coast." *Journal of Latin American and Caribbean Anthropology* 15 (1): 171–94.

Brondo, Keri V. 2011. "From Fishing to Tourism? Conservation, Development, and Garifuna Activism in the Cayos Cochinos Marine Protected Area." *Cengage Learning's Anthropology Course Reader*. Detroit: Gale.

Brondo, Keri V. 2013. *Land Grab: Green Neoliberalism, Gender, and Garifuna Resistance in Honduras*. Tucson: University of Arizona Press.

Brondo, Keri V. 2015. "The Spectacle of Saving: Conservation Voluntourism and the New Neoliberal Economy on Utila, Honduras." *Journal of Sustainable Tourism* 23 (10): 1405–25.

Brondo, Keri V. 2018. "'A Dot on a Map': Gender, Communal Land Titling, and the Paradoxes of Indigenous Rights for Honduran Garifuna." *Political and Legal Anthropology* 41 (2): 185–200.

Brondo, Keri V. 2019. "Entanglements in Conservation: Utila's Emerging Economy of Affect." *Journal of Sustainable Tourism* 27 (4): 590–627.

Brondo, Keri, and Natalie Bown. 2011. "Neoliberal Conservation, Garifuna Territorial Rights and Resource Management in the Cayos Cochinos Marine Protected Area." *Conservation and Society* 9 (2): 91–105.

Burridge, M. J. 2005. "Controlling and Eradicating Tick Infestations on Reptiles." *Parasitology* 27 (5): 371–76.

Büscher, Bram, and William Dressler. 2007. "Linking Neoprotectionism and Environmental Governance: On the Rapidly Increasing Tensions Between Actors in the Environment-Development Nexus." *Conservation and Society* 5 (4): 586–611.

Büscher, Bram, William Dressler, and Robert Fletcher. 2014. *Nature™ Inc.: New Frontiers of Environmental Conservation in the Neoliberal Age*. Tucson: University of Arizona Press.

Büscher, B., S. Sullivan, K. Neves, J. Igoe, and D. Brockington. 2012. "Towards A Synthesized Critique of Neoliberal Biodiversity Conservation." *Capitalism, Nature, Socialism* 23 (2): 4–30. https://doi.org/10.1080/10455752.2012.674149.

Butcher, Jim. 2003. *The Moralisation of Tourism: Sun, Sand . . . and Saving the World?* New York: Routledge.

Butcher, Jim. 2011. "Volunteer Tourism May Not Be as Good as It Seems." *Tourism and Recreation Research* 36 (1): 74–76.

Butcher, Jim, and Peter Smith. 2010. "'Making a Difference': Volunteer Tourism and Development." *Tourism Recreation Research* 35 (1): 27–36.

Bryman, Alan. 2004. *The Disneyization of Society*. London: Sage.

Bryman, Alan. 2009. "The Disneyization of Society." In *Service Work: Critical Perspectives*, edited by Cameron L. Macdonald and Marek Korczynski, 53–72. New York: Routledge.

Caissie, L. T., and E. A. Halpenny. 2003. "Volunteering for Nature: Motivations for Participating in a Biodiversity Conservation Volunteer Program." *World Leisure Journal* 45 (2): 38–50. https://doi.org/10.1080/04419057.2003.9674315.

Canty, Steven W. J. 2007. "Positive and Negative Impacts of Dive Tourism: The Case Study of Utila, Honduras." MA thesis. Lund University Centre for Sustainability Studies, Sweden.

Carrier, James, G. and Donald V. L. Macleod. 2005. "Bursting the Bubble: The Sociocultural Context of Ecotourism." *Journal of the Royal Anthropological Institute* 11 (2): 315–34.

Chrulew, Matthew, Stuart Cooke, Matthew Kearnes, Emily O'Gorman, Deborah Rose, and Thom van Dooren. 2012. "Thinking Through the Environment, Unsettling the Humanities." *Environmental Humanities* 1 (1): 1–5.

Coghlan, Alexandra. 2007. "Towards an Integrated Image-Based Typology of Volunteer Tourism Organisations." *Journal of Sustainable Tourism* 15 (3): 267–87. https://doi.org/10.2167/jost628.0.

Collard, Rosemary-Claire, Jessica Dempsey, and Juanita Sundberg. 2015. "A Manifesto for Abundant Futures." *Annals of the Association of American Geographers* 105 (2): 322–30.

Crick, Malcolm. 1995. "The Anthropologist as Tourist: An Identity in Question." In *International Tourism: Identity and Change*, edited by M. F. Lanfant, J. B. Allcock, and E. M. Bruner, 205–23. London: Sage Publications.

Cruzen, P. J., and E. F. Stoermer. 2000. "The 'Anthropocene.'" *Global Change Newsletter* 41:17–18.

Currin, Frances Heyward. 2002. "Transformation of Paradise: Geographical Perspectives on Tourism Development on a Small Caribbean Island (Utila, Honduras)." MA thesis, Louisiana State University and Agricultural and Mechanical College.

Da Silva, Denise Ferreira. 2015. "Before Man: Sylvia Wynter's Rewriting of the Modern Episteme." In *Sylvia Wynter: On Being Human as Praxis*, edited by Katherine McKittrick, 90–105. Durham, N.C.: Duke University Press.

Davidson, William V. 1974. *Historical Geography of the Bay Islands, Honduras: Anglo-Hispanic Conflict in the Western Caribbean*. Birmingham, Ala.: Southern University Press.

Debord, Guy. (1967) 1995. *Society as Spectacle*. New York: Zone.

Defoe, Daniel. (1719) 2007. *Robinson Crusoe*. Oxford: Oxford University Press.

Deumert, Ana. 2019. "The Rhizome: Multiplicity and Assemblage." *Diggit Magazine*. https://www.diggitmagazine.com/column/mangrove-or-moving-and-beyond-rhizome.

Devine, Jennifer. 2014. "Counterinsurgency Ecotourism in Guatemala's Maya Biosphere Reserve." *Environment and Planning D: Society and Space* 32(6): 984–1001.

Douglas, Jason A. 2014. "What's Political Ecology Got to Do with Tourism?" *Tourism Geographies* 16(1): 8–13.

Driver, Alice. "Utila Honduras: Killing Lionfish to Save the Seas." Go NOMAD. Last accessed February 12. 2020. https://www.gonomad.com/3163-utila-honduras-killing-lionfish-save-seas/.

Dubois, W. E. B. 1915. "The African Roots of War." *Atlantic Monthly*, May.

Dupuis, John. 2017. "A Visit to Famous Dr. John in Utila." *Honduras Travel*. https://hondurastravel.com/featured/visit-famous-dr-john-utila/.

Escobar, Arturo. 2017. *Designs for the Pluriverse: Radical Interdependence, Autonomy, and the Making of Worlds*. Durham, N.C.: Duke University Press.

Fassin, Didier. 2011. *Humanitarian Reason: A Moral History of the Present*. Berkeley: University of California Press.

Feldman, Ilana, and Miriam Ticktin. 2010. *In the Name of Humanity: The Government of Threat and Care*. Durham, N.C.: Duke University Press.

Feldman, Joseph P. 2011. "Producing and Consuming 'Unspoilt' Tobago: Paradise Discourse and Cultural Tourism in the Caribbean." *Journal of Latin American and Caribbean Anthropology* 16 (1): 46–66.

Field Notes from Utila. July 15–August 7, 2016.

Figueroa, Alejandro J., Whitney. A. Goodwin, and E. Christian Wells. 2012. "Mayanizing Tourism on Roatán Island, Honduras: Archaeological Perspectives on Heritage, Development, and Indigeneity." In *Global Tourism: Cultural Heritage and Economic Encounters*, edited by S. Lyon and F. Wells, 43–60. Lanham, Md.: Altamira Press.

Fisher, William. 1997. "Doing Good? The Politics and Anti-politics of NGO Practices." *Annual Review of Anthropology* 26:439–64.

Fletcher, Robert. 2010. "Neoliberal Environmentality: Towards a Poststructuralism Political Ecology of the Conservation Debate." *Conservation and Society* 8 (3): 171–81.

Fletcher, Robert, William Dressler, and Bram Büscher. 2014. "Introduction." In *Nature™ Inc.: New Frontiers of Environmental Conservation in the Neoliberal Age*, edited by Bram Büscher, Wolfram Dressler, and Robert Fletcher, 3–24. Tucson: University of Arizona Press.

Foucault, Michel. 2008. *The Birth of Biopolitics: Lectures at the College de France, 1978–1979*. New York: Palgrave.

Freidus, Andrea. 2010. "'Saving' Malawi: FAITHFUL Responses to Orphans and Vulnerable Children." *North American Practicing Anthropology Bulletin* 33 (1): 50–67.

Freidus, Andrea. 2017. "Unanticipated Outcomes of Voluntourism Among Malawi's Orphans." *Journal of Sustainable Tourism* 25 (9): 1–16. https://doi.org/10.1080/09669582.2016.1263308.

Glissant, Édouard. 1990. *Poetics of Relation*. Translated by Betsy Wing. Ann Arbor: University of Michigan Press.

Gmelch, Sharon. B. 2010. "Why Tourism Matters." In *Tourists and Tourism, A Reader*. 2nd ed, edited by Sharon B. Gmelch, 3–24. Long Grove, Ill.: Waveland Press.

Graburn, Nelson H. H. 2010. "Secular Ritual: A General Theory of Tourism." In *Tourists and Tourism: A Reader*, 2nd ed., edited by Sharon B. Gmelch, 25–56. Long Grove, Ill.: Waveland Press.

Graburn, Nelson. 2012. "Tourism: The Sacred Journey." In *Hosts and Guests. The Anthropology of Tourism*, edited by Valene L. Smith, 21–36. Philadelphia: University of Pennsylvania Press.

Graham, Ross. 2010. "Honduras / Bay Island English." In *The Lesser-Known Varieties of English: An Introduction*, edited by Daniel Schreier, Peter Trudgill, Edgar W. Schneider, and Jeffrey P. Williams, 92–135. Cambridge: Cambridge University Press.

Grandia, Liza. 2007. "Between Bolivar and Bureaucracy: The Mesoamerican Biological Corridor." *Conservation and Society* 5 (4): 478–503.

Guttentag, Daniel. 2009. "The Possible Negative Impacts of Volunteer Tourism." *International Journal of Tourism Research* 11 (6): 537–51. https://doi.org/10.1002/jtr.727.

Guttentag, Daniel. 2010. "Volunteer Tourism: As Good as It Seems?" *Tourism Recreation Research* 36(1): 69–74.

Hardt, Michael. 1999. "Affective Labour." *Boundary* 26 (2): 89–100.

Haraway, Donna. 2008. *When Species Meet*. Minneapolis: University of Minnesota Press.

Haraway, Donna. 2016. *Staying with the Trouble: Making Kin in the Chthulucene*. Durham, N.C.: Duke University Press.

Harborne, Alastair R., Daniel C. Afzal, and Mark J. Andrews. 2001. "Honduras: Caribbean Coast." *Marine Pollution Bulletin* 42 (12): 1221–35.

Harcourt, Wendy, and Ingrid Nelson, eds. 2015. *Practicing Feminist Political Ecology: Moving Beyond the Green Economy*. London: Zed Books.

Heath, Sue. 2007. "Widening the Gap: Pre-university Gap Years and the 'Economy of Experience.'" *British Journal of Sociology of Education* 28 (1): 89–103.

Hoffman, David M. 2014. "Conch, Cooperatives, and Conflict: Conservation and Resistance in the Banco Chinchorro Biosphere Reserve." *Conservation and Society* 12 (2): 120–32.

Holmes, George, and Connor Joseph Cavanagh. 2016. "A Review of the Social Impacts of Neoliberal Conservation: Formations, Inequalities, Contestations." *Geoforum* 75: 199–209.

Igoe, Jim. 2013. "Nature on the Move II: Contemplation Becomes Speculation." *New Proposals: Journal of Marxism and Interdisciplinary Inquiry* 6 (1–2): 37–49.

Igoe, Jim. 2016. *The Nature of Spectacle: On Image, Money, and Conserving Capitalism*. Tucson: University of Arizona Press.

Igoe, Jim. 2017. "Afterword." In *Political Ecology of Tourism: Community, Power and the Environment*, edited by M. Mostafanezhad, R. Norum, E. J. Shelton, and A. Thompson-Carr, 309–16. London: Routledge.

Igoe, Jim., and D. Brockington. 2007. "Neoliberal Conservation: A Brief Introduction." *Conservation and Society* 5 (4): 432–499.

Igoe, Jim., K. Neves, and D. Brockington. 2010. "A Spectacular Eco-Tour around the Historic Bloc: Theorising the Convergence of Biodiversity Conservation and Capitalist Expansion." *Antipode* 42 (3): 486–512.

Igoe, Jim, and Sian Sullivan. 2008. "Problematizing Neoliberal Biodiversity Conservation: Displaced and Disobedient Knowledges, An Executive Summary for the International Institute for Environment and Development." ResearchGate. https://www.researchgate.net/publication/291835496_Problematizing_neoliberal_biodiversity_conservation_displaced_and_disobedient_knowledges.

"Information about Sea Turtles: Threats from Marine Debris," n.d. Sea Turtle Conservancy. https://conserveturtles.org/information-sea-turtles-threats-marine-debris/.

Jackson, Sue, and Lisa R. Palmer. 2014. "Reconceptualizing Ecosystem Services Possibilities for Cultivating and Valuing the Ethics and Practices of Care." *Progress in Human Geography* 39(2): 122–45.

Jackson, William. 2003. *And the Sea Shall Hide Them*. Milwaukee: Gardenia Press.

Johnson, Melissa A. 2018. *Becoming Creole: Nature and Race in Belize*. New Brunswick, N.J.: Rutgers University Press.

Kent, Suzanne, and Keri Vacanti Brondo. 2019. "'Years Ago the Crabs Was So Plenty': Anthropology's Role in Ecological Grieving and Conservation Work." *Culture, Agriculture, Food and Environment*. https://doi.org/10.1111/cuag.12235.

Kirksey, Eben. 2012. *Freedom in Entangled Worlds: West Papua and the Architecture of Global Power*. Durham, N.C.: Duke University Press.

Kirksey, Eben. 2014. *The Multispecies Salon*. Durham, N.C.: Duke University Press.

Kirksey, Eben. 2015. *Emergent Ecologies*. Durham, N.C.: Duke University Press.

Kohn, Eduardo. 2007. "How Dogs Dream: Amazonian Natures and the Politics of Transspecies Engagement." *American Ethnologist* 34 (1): 3–24.
Kohn, Eduardo. 2013. *How Forests Think: Toward an Anthropology Beyond the Human*. Berkeley: University of California Press.
Latour, Bruno. 2004. "How to Talk About the Body? The Normative Dimension of Science Studies." *Body and Society* 10 (2–3): 205–29.
Latour, Bruno. 2007. *Reassembling the Social*. Oxford: Oxford University Press.
Law, John. 2015. What's Wrong with a One-World World? *Distinktion: Journal of Social Theory* 16 (1): 126–39.
Le Guin, Ursula K. 2004. "Telling Is Listening." In *The Wave in the Mind: Talks and Essays on the Writer, the Reader, and the Imagination*, 156–172. Boston, Mass.: Shambhala Publications.
Little, Becky. "Debunking the Myth of the 'Real' Robinson Crusoe." n.d. National Geographic online. https://www.nationalgeographic.com/news/2016/09/robinson-crusoe-alexander-selkirk-history/#close.
Lord, David. 1975. "Money Order Economy: Remittances in the Island of Utila." PhD diss., University of California Riverside.
Luxemburg, Rosa. 2003. *The Accumulation of Capital*. Translated by Agnes Schwarzschild. London: Routledge Classics.
Lyons, Kevin, Joan Hanley, Stephen Wearing, and John Neil. 2012. "Gap Year Volunteer Tourism: Myths of Global Citizenship?" *Annals of Tourism Research* 39 (1): 361–78.
Lyons, Kevin, and Stephen Wearing. 2008. "The Volunteer's Journey Through Leisure into the Self." In *Journeys of Discovery in Volunteer Tourism*, edited by K. Lyon and S. Wearing, 63–71. Cambridge, Mass.: CABI Publishing.
MacCannell, Dean. 1971. *The Tourist: A New Theory of the Leisure Class*. New York: Schocken Books.
Malkki, Liisa. 1996. "Speechless Emissaries: Refugees, Humanitarianism, and Dehistoricization." *Cultural Anthropology* 11 (3): 377–404.
Malkki, Liisa. 2010. *In the Name of Humanity: The Government of Threat and Care*. Durham, N.C.: Duke University Press.
Mangrove Crabs. 2021. "Mangrove Crabs: Types, Main Characteristics and Importance." Our Marine Species. http://ourmarinespecies.com/c-crabs/mangrove-crabs/.
MartyandGinski. 2013. "If You Come to Utila." August 4. YouTube video. https://www.youtube.com/watch?v=LX-H-zF9PJ4.
Marx, Karl (1867) 1957. *Capital*. Vol. 1. London: J. M. Dent and Sons Ltd.
Maryon, Daisy, Stesha Pasachnik, Diego Ardón, and Steve Clayson. 2018. "*Ctenosaura bakeri* Redlist Assessment." ResearchGate. https://www.researchgate.net/publication/326357422_Ctenosaura_bakeri_Redlist_Assessment_2018.
IUCN. 2018. "The IUCN Red List of Threatened Species 2018." RedList. Retrieved on February 17, 2020. https://doi.org/10.2305/iucn.uk.2018-1.rlts.t44181a125203850.en.

Maturana, Humberto, and Gerda Verden-Zöller. 2008. *The Origin of Humanness in the Biology of Love*. Charlottesville, Va.: Imprint Academic.
McGehee, Nancy G. 2002. "Alternative Tourism and Social Movement." *Annals of Tourism Research* 29 (1): 124–43. https://doi.org/10.1016/S0160-7383(01)00027-5.
McGehee, Nancy G. 2012. "Oppression, Emancipation, and Volunteer Tourism: Research Propositions." *Annals of Tourism Research* 39 (1): 84–107.
McGehee, Nancy G. 2014. "Volunteer Tourism: Evolution, Issues, and Futures." *Journal of Sustainable Tourism* 22 (6): 847–54.
McGehee, Nancy G., and W. C. Norman. 2002. "Alternative Tourism as Impetus for Consciousness-Raising." *Tourism Analysis* 6 (3–4): 239–51. https://doi.org/10.3727/108354201108749872.
McGehee, Nancy G., and C. A. Santos. 2005. "Social Change, Discourse, and Volunteer Tourism." *Annals of Tourism Research* 32 (3): 760–79.
McNab, Shelby. n.d. *The True Story of Robinson Crusoe*. Pamphlet and Tour Guide Maps. Produced and edited by Shirley Canther. Deltona, Fla.: TLC Innovations.
Mendoza, Marcos. 2018. *The Patagonian Sublime: The Green Economy and Postneoliberal Politics*. New Brunswick, N.J.: Rutgers University Press.
Moore, Amelia. 2012. "The Aquatic Invaders: Marine Management Figuring Fishermen, Fisheries, and Lionfish in the Bahamas." *Cultural Anthropology* 27 (4): 667–88.
Moore, Amelia. 2019. *Destination Anthropocene: Science and Tourism in the Bahamas*. Oakland: University of California Press.
Mostafanezhad, Mary. 2013. "'Getting in Touch with Your Inner Angelina': Celebrity Humanitarianism and the Cultural Politics of Gendered Generosity in Volunteer Tourism." *Third World Quarterly* 34 (3): 486–99.
Mostafanezhad, Mary, Roger Norum, Eric J. Shelton, and Anna Thompson-Carr. 2017. "Introduction." In *Political Ecology of Tourism: Community, Power and the Environment*, edited by M. Mostafanezhad, R. Norum, E. J. Shelton, and A. Thompson-Carr, 1–21. London: Routledge.
Neves, Katja. 2005. "Chasing Whales with Bateson and Daniel." *Australian Humanities Review* 35. http://australianhumanitiesreview.org/2005/06/01/chasing-whales-with-bateson-and-daniel/?utm_source=rss&utm_medium=rss&utm_campaign=chasing-whales-with-bateson-and-daniel/.
Neves, Katja. 2010. "Cashing in on Cetourism: A Critical Ecological Engagement with Dominant E-NGO Discourses on Whaling, Cetacean Conservation, and Whale Watching." *Antipode* 42 (3): 719–41.
Novakova, Marketa, Ivan Literak, Luis Chevez, Thiago F. Martins, Maria Ogrzewalska, and Marcelo B. Labruna. 2015. "Rickettsial Infections in Ticks from Reptiles, Birds and Humans in Honduras." *Ticks and Tick Borne Diseases* 6 (6): 737–42. https://www.sciencedirect.com/journal/ticks-and-tick-borne-diseases/vol/6/issue/6.
Ogden, Laura A. 2011. *Swamplife: People, Gators, and Mangroves Entangled in the Everglades*. Minneapolis: University of Minnesota Press.

Ogden, Laura A., Billy Hall, and Kimiko Tanita. 2013. "Animals, Plants, People, and Things: A Review of Multispecies Ethnography." *Environment and Society* 4(1): 5–24.

Ong, Aihwa, and Stephen Collier, eds. 2005. *Global Assemblages*. Hoboken, N.J.: Wiley-Blackwell.

Ooi, Natalie, and Jennifer H. Laing. 2010. "Backpacker Tourism: Sustainable and Purposeful?: Investigating the Overlap Between Backpacker Tourism and Volunteer Tourism Motivations." *Journal of Sustainable Tourism* 18 (2): 191–206.

Palacios, Carlos M. 2010. "Volunteer Tourism, Development and Education in a Postcolonial World: Conceiving Global Connections Beyond Aid." *Anatolia: An International Journal of Tourism and Hospitality Research* 20 (1): 861–78. https://doi.org/10.1080/09669581003782739.

Parreñas, Rheana "Juno" Salazar. 2012. "Producing Affect: Transnational Volunteerism in a Malaysian Orangutan Rehabilitation Center." *American Ethnologist* 39 (4): 673–87.

Peet, Richard, Paul Robbins, and Michael Watts, eds. 2011. *Global Political Ecology*. London: Routledge.

Perley, Bernard, ed. Forthcoming. *Remediating Cartographies of Erasure: Anthropology, Indigenous Epistemologies, and the Global Imaginary*. Lincoln: University of Nebraska Press.

Pile, Steve. 2010. "Emotions and Affect in Recent Human Geography." *Transactions of the Institute of British Geographers, New Series* 35 (1): 5–20.

Pülmanns Nathalie, Ulf Mehlig, Inga Nordhaus, Ulrih Saint-Paul, and Karen Diele. 2016. "Mangrove Crab *Ucides cordatus* Removal Does Not Affect Sediment Parameters and Stipule Production in a One Year Experiment in Northern Brazil." *PLoS ONE* 11 (12): e0167375. https://doi.org/10.1371/journal.pone.0167375.

Rahier, Jean Muteba. 2012. *Black Social Movements in Latin America: From Monocultural Mestizaje to Multiculturalism*. London: Palgrave Macmillian.

Raymond, Eliza Marguerite, and C. Michael Hall. 2008. "The Development of Cross-Cultural (Mis)Understanding Through Volunteer Tourism." *Journal of Sustainable Tourism* 16 (5): 530–43.

Robbins, Paul. 2019. *Political Ecology: A Critical Introduction*, 3rd ed. Hoboken, N.J.: Wiley-Blackwell.

Rocheleau, Diane, Barbara Thomas-Slayter, and Esther Wangari. 2013. *Feminist Political Ecology: Global Issues and Perspectives*. New York: Routledge.

Rose, Richard H. 1904. *Utilla: Past and Present*. Dansville, N.Y.: F.A. Owen Publishing Company.

Roth, Robin J., and Wolfram Dressler. 2012. Market-Oriented Conservation Governance: The Particularities of Place. *Geoforum* 43 (3): 363–66. https://doi.org/10.1016/J.geoforum.2012.01.006.

Ruddick, Susan. 2010. "The Politics of Affect: Spinoza in the Work of Negri and Deleuze." *Theory, Culture, and Society* 27(4): 21–45.

Salazar, Noel B. 2012. "Tourism Imaginaries: A Conceptual Approach." *Annals of Tourism Research* 39(2): 863–82.
Salazar, Noel B., and Nelson H. H. Graburn. 2014. "Introduction: Toward an Anthropology of Tourism Imaginaries." In *Tourism Imaginaries: Anthropological Approaches*, edited by Noel B. Salazar and Nelson H. H. Graburn, 1–30. New York: Berghahn Press.
Sauer, Carl Ortwin. 1966. *The Early Spanish Main*. Cambridge: Cambridge University Press.
Schuller, Mark. 2009. "Gluing Globalization: NGOs as Intermediaries in Haiti." *Polar: Political and Legal Anthropology Review* 32 (1): 84–104.
Schulz, Karsten A. 2017. "Decolonizing Political Ecology: Ontology, Technology, and 'Critical Enchantment.'" *Journal of Political Ecology* 24(1): 125–43.
Schumpeter, Joseph. 1942. *Capitalism, Socialism, and Democracy*. London: Routledge.
Sheller, Mimi. 2003. *Consuming the Caribbean: From Arawaks to Zombies*. London: Routledge.
Simpson, Kate. 2004. "'Doing Development': The Gap Year, Volunteer-Tourists and a Popular Practice of Development." *Journal of International Development* 16 (5): 681–92. https://doi.org/10.1002/jid.1120.
Simpson, Kate. 2005. "Broad Horizons? Geographies and Pedagogies of the Gap Year." PhD diss., Newcastle University, Newcastle Upon Tyne, Tyne and Wear, UK.
Singh, Neera. 2013. "The Affective Labor of Growing Forests and the Becoming of Environmental Subjects: Rethinking Environmentality in Odisha, India." *Geoforum* 47 (June): 189–98.
Skoggard, Ian, and Alisse Waterston. 2015. "Introduction: Toward an Anthropology of Affect and Evocative Ethnography." *Anthropology of Consciousness* 26(2): 109–20. https://doi.org/10.1111/anoc.12041.
Smedley, Audrey. 1993. *Race in North America: Origin and Evolution of a Worldview*. Boulder, Colo.: Westview Press.
Smith, Artly Emile Brooks. 2013. *Black Chest: A Vault of Historical and Cultural Knowledge Stored by the Black English Speaking People That Reside in the Bay Islands and the Honduran North Coast*. Tegucigalpa, Honduras: Native Bay Islanders Professional and Laborers Association Press (NABIPLA). Ediciones Guardabarranco.
Söderman, N., and S. L. Snead. 2008. "Opening the Gap: The Motivation of Gap Year Travelers to Volunteer in Latin America." In *Journeys of Discovery in Volunteer Tourism*, edited by K. Lyons and S. Wearing. 118–29. Wallingford, UK: CABI Publishing.
Stonich, Susan C. 1998. "Political Ecology of Tourism." *Annals of Tourism Research* 25(1): 25–54.
Stonich, Susan C. 1999. *The Other Side of Paradise: Tourism, Conservation, and Development in the Bay Islands*. New York: Cognizant Communication Corporation.
Strachan, Ian G. 2002. *Paradise and Plantation: Tourism and Culture in the Anglophone Caribbean*. Charlottesville: University of Virginia Press.

Strong, William Duncan. 1935. "Archaeological Investigations in the Bay Islands, Spanish Honduras." *Smithsonian Miscellaneous Collections*. Vol. 92, no. 14, February 12. Washington, D.C.: Smithsonian Institution.

Sullivan, Sian. 2013. "Banking Nature?: The Spectacular Financialisation of Environmental Conservation." *Antipode* 45 (1): 198–217.

Sullivan, Sian, Jim Igoe, and Bram Büscher. 2013. "Introducing 'Nature on the Move'—A Triptych." *Proposals: Journal of Marxism and Interdisciplinary Inquiry* 6 (1–2): 15–19.

Sundberg, Juanita. 2013. "Decolonizing Posthumanist Geographies." *Cultural Geographies* 21 (1): 33–47.

Sutcliffe, Joe. 2012. "A Critical Perspective on Volunteer Tourism and Development." E-International Relations. https://www.e-ir.info/2012/10/04/international-citizen-service-a-critical-perspective-on-volunteer-tourism-and-development/.

Ticktin, Miriam. 2011. *Causalities of Care: Immigration and the Politics of Humanitarianism in France*. Oakland: University of California Press.

Tomazos, Kostas, and Richard Butler. 2009. "Volunteer Tourism: The New Ecotourism?" *Anatolia: An International Journal of Tourism and Hospitality Research* 20 (1): 196–212.

Tompson, Jon. 2019. "The Paya Resistance." *Päyä: The Roatan Lifestyle Magazine*. https://payamag.com/2019/12/20/the-paya-resistance/.

Tsing, Anna L. 2012. "Unruly Edges: Mushrooms as Companion Species." *Environmental Humanities* 1(1): 141–54.

Tsing, Anna L. 2015. *The Mushroom at the End of the World: On the Possibility of Life in Capitalist Ruins*. Princeton, N.J.: Princeton University Press.

Turner, Victor. 1967. "Betwixt and Between: The Liminal Period in Rites de Passage." *Forest of Symbols: Aspects of the Ndembu Ritual*, 23–59. Ithaca, N.Y.: Cornell University Press.

Turner, Victor. 1969. *The Ritual Process*. Harmondsworth, UK: Penguin Books.

Turner, Victor, and Edith L. B. Turner. 1978. *Image and Pilgrimage in Christian Culture: Anthropological Perspectives*. New York: Columbia University Press.

Utila Lodge. 2017. Dive packages (have since changed). https://utilalodge.com.

Utila Lobster Diver Program. 2018. GoFundMe Campaign. https://www.gofundme.com/f/utila-cays-lobster-divers.

Van Gennep, Arnold. (1909) 1960. *The Rites of Passage*. Translated by Monika Vizedom and Gabrielle Caffee. Chicago: University of Chicago Press.

Van Landeghem, M. 2017. "When You're Aware, You Care." *WSORC* (blog), April 7. Last accessed November 17, 2017. https://wsorc.org/when-youre-aware-you-care/.

Vodopivec, Barbara, and Rivke Jaffe. 2011. "Save the World in a Week: Volunteer Tourism, Development and Difference." *European Journal of Development Research* 23(1): 111–28.

Volunteer Forever. n.d. "Original Volunteers." Last accessed March 20, 2021. volunteerforever. https://www.volunteerforever.com/program/original-volunteers/.

Vrasti, Wanda. 2012. *Volunteer Tourism in the Global South: Giving Back in Neoliberal Times*. New York: Routledge.
Wade, Peter. 2003. "Race and Nation in Latin America: An Anthropological View." In *Race and Nation in Modern Latin America*, edited by N. P. Appelbaum, A. S. Macpherson, and K. A. Rosemblatt, 263–81. Chapel Hill: University of North Carolina Press.
Wade, Peter. 2010. *Race and Ethnicity in Latin America*. 2nd ed. London: Pluto Press.
Waitt, Gordon R., Robert Figueroa, and Tom Nagle. 2014. "Paying for Proximity: Touching the Moral Economy of Ecological Tourism." In *Moral Encounters in Tourism*, edited by M. Mostafanezhad and K. Hannam, 167–81. London: Ashgate Publishing.
Watts, Michael, and Richard Peet. 2004. "Liberating Political Ecology." In *Liberation Ecologies: Environment, Development and Social Movements*, edited by R. Peet and M. Watts, 3–47. New York: Routledge.
Wearing, Stephen. 2001. *Volunteer Tourism: Experiences That Make a Difference*. Wallingford, UK: CABI International.
Wearing, Stephen. 2002. "Re-centering the Self in Volunteer Tourism." In *The Tourist as a Metaphor of the Social World*, edited by G. S. Dann, 237–62. Wallingford, UK: CABI.
Wearing, Stephen, A. Deville, and Kevin Lyons. 2008. "The Volunteer's Journey Through Leisure into the Self." In *Journeys of Discovery in Volunteer Tourism*, edited by K. Lyon and S. Wearing, 63–71. Cambridge, Mass.: CABI Publishing.
Wearing, Stephen, and Nancy McGehee. 2013a. "Volunteer Tourism: A Review." *Tourism Management* 38 (October): 120–30. https://doi.org/10.1016/j.tourman.2013.03.002.
Wearing, Stephen, and Nancy G. McGehee. 2013b. *International Volunteer Tourism: Research, Theory, and Practice*. Wallingford, UK: CABI.
Wearing, Stephen, and Michael Wearing. 2006. "'Rereading the Subjugating Tourist' in Neoliberalism: Postcolonial Otherness and the Tourist Experience." *Tourism Analysis* 11 (2): 145–63.
Wells, Christian E. 2008. "La arqueología y el futuro del pasado en las Islas de la Bahía." *Yaxkin* 24 (1): 66–81.
West, Paige. 2016. *Dispossession and the Environment: Rhetoric and Inequality in Papua New Guinea*. New York: Columbia University Press.
West, Paige, J. Igoe, and Dan Brockington, 2006. "Parks and Peoples: The Social Impacts of Protected Areas." *Annual Review of Anthropology* 35:251–77.
"What Is Paya Magazine." 2020. *Päyä: The Roatan Lifestyle Magazine*. https://payamag.com/about-us/what-is-paya-magazine/.
Wynter, Sylvia. 2003. "Unsettling the Coloniality of Being/Power/Truth/Freedom: Towards the Human, After Man, Its Overrepresentation—An Argument." *CR: The New Centennial Review* 3 (3): 257–337.

Index

Page numbers in *italics* refer to figures.

"80 acre" residential area of Utila, 44

AboutUtila.com, 76
affect: assemblages of, 117–18; between orchid bee and orchid, 6; in capitalism, 25; decolonial feminist scholars on, 167–68; and encounter with other species, 18–20, 31–32, 95, 146, 149–50, 164, 170; in food traditions on Utila, 37–38, 138, 164; in multispecies conservation voluntourism, 33, 127, 138, 148–53, 154–55, 171n6; political ecology on, 31–32, 166; in stewardship science, 146; and subjectivities, 147; in water, 156
affect economy: conservation voluntourism in, 116–18; 138, 140, 148–53, 164–66; dispossession in, 132; exchange value in, 24–25, 116, 149, 155, 164–66, 168–69, 172n14; life and death of species in, 20, 33, 140, 157, 170; material outcomes in, 138; tourist imaginaries in, 163–64; of Utila, 37–38, 69, 138, 140, 163, 167–70

affective labor: and "becoming with," 166; and engagement with other beings, 164–66; exchange value of, 116, 149, 155, 164–6, 168–69, 172n14; and "more-than-human," 8, 111; transforming subjectivities, 14, 147; voluntourism as, 22–27, 116, 140–41, 148–53, 154–55, 164–66, 171n6
"affective turn" in social sciences, 167–68
African beliefs, survival of, 40–41
African origins, in Utila stratification, 67–68
Afro-Antillean people, and slavery, 9–10
Afro-Brazilian traditions, 64–65
agency: of diverse beings, 21, 168–69; of nonhumans, 24, 61, 65, 168, 170; and think-with the-mangrove, 65
agriculture: boom in, 12; decline of, 14; dispossession of, 149; subsistence level, 9; turtles and, 37
"Alex in Wonderland" blog, 87–88
Alice (herpetologist), 6, 157–63
Amalie (daughter of author), 89, 157–63

American Museum of Natural History, 42
Amerindians, 9–10
And the Sea Shall Hide Them (Jackson), 53–63, 64, 65–66, 162–63
Anglo-Antillean people, and slavery, 9–10
Anthony Keys Resort decompression station, 142
Anthropocene era: human activities affecting planet in, 113, 148; island places and, 123; life beyond, 166; "more-than-human" in, 20–22; moving beyond, 168, 170; urgencies of, 167
Anthropocene tourism, conservation voluntourism as, 113
anthropology: on assemblages, 52, 114; on humanitarianism, 117; and multispecies encounters, 114–15; on tourism, 114, 177n2
Arawakan Indigenous people, 9–10
archaeology of Utila, 42–46
assemblages: Deumert and Ogden on, 22; and imaginaries, 135; in multispecies relationality, 8, 22, 52, 64, 69, 111, 138, 149–53, 169–70;
authenticity, as tourist motivation, 73

Bambu area of Utila, 44, 45–46
bananas: plantations growing, 9–10, 12; U.S. fruit companies and, 9, 12, 13
Bando Beach, development at, 104–106
Bando Beach (restaurant), 46
bats, cave habitat of, 40
Bay Islands: ethnicities in, 9–11; history of, 45–50, 52–54; oral culture of, 40–41; as Tourism Free Zone, 95; waters around, 17
Bay Islands College of Diving (BICD): hyperbaric chamber at, 141; and WSORC, 133–34
Bay Islands Colony, 12–13, 50, 52–54
Bay Islands Conservation Association (BICA): dive buoys project, 119–20, 141; founders of, 118–19, 121–23; in lionfish collaboration, 18, 138; marine projects of, 17–18; as NGO, 16; as protected area comanager, 1; Reef Leaders program, 122, 128, 178n11; spearfishing training by, 138; staff turnover at, 134–35, 156; and tourism economy, 100, 119; turtle projects of, 37–39, 120–22, 156, 163, 174n2; volunteer experiences at, 18, 155–56, 163, 179n21
"Bay Islands English" (BIE), 11–12, 40–41
Bay Islands Marine Park, 17, 173n4
beaches: cleaning by volunteers, 155–56; coconuts in environment of, 70, 72, 96–97; construction reducing, 39, 103–6, 148–49, 173n4; privatization of, 105–6; protection of, 69, 118–19; sea turtle nesting on, 15, 39, 69, 155, 173n4
Becoming Creole (Johnson), 51, 175n9
"becoming with": affective labor in, 19, 166; as goal of conservation voluntourism, 152; Haraway on, 8, 167; through multispecies engagement, 116–17, 148, 149, 166
"biology of love," 30, 169
Bird, Junius, 42
Birth of Biopolitics, The (Foucault), 133–31
Black Chest (Smith), 40–41, 53
black mangrove (*Avicennia germinans*): as swamper iguana habitat, 14–15, 128; propagules of, 180n2; salt tolerance of, 161; stench of, 161–62. *See also* mangroves
blackness and brownness, encoded ideas on, 51
Black Rock Basin burial site, 45
blanquimiento process of "whitening," 10
Blaser, Mario, on political ontology, 21–22
Blueberry Hill (boardinghouse), 85
blue crab: harvesting regulated, 8, 127–28; in historical fiction, 61–62; iguanas eating of, 127, 169; and mangrove, 62, 127; as nocturnal, 62–62, 76; and raccoons, 168–69, 170

boas: conservation of, 153–55; and ticks, 153–54
bonito tuna: drawn by whale shark activity, 112, 115, 133; scarcity due to whale shark tourism, 136–37
Brandon Hill Cave: as bat habitat, 40; as "Robinson Crusoe Cave, 40, 42; stories of artifacts found in, 41–42
British Empire: Bay Islands Colony under, 12–13, 50, 52–54; conflict with Spain, 9, 31, 34, 49–50, 52–53; Utilian heritage and, 12, 13
Bryman, Alan, on "Disneyization," 164–65
Bucket of Blood (bar), 85
Büscher, Bram, on environmental impact of tourism, 23
"bush," the: as zone of freedom, 71–72, 176n3
buttonwood mangrove (*Conocarpus erectus*), 14–15, 161–62. See also mangroves

Callejas, Rafael, administration of, 105
Camponado neighborhood: crime in, 68; as former wetland, 35; low-cost housing development in, 106–9, *107*; as shanty development 36, *36*, 62–63
capital: cultural, 19–20, 118, 130, 132, 138, 173n16; and dispossession, 109–10; exchange of, 19–20, 75, 116, 118, 138, 149; natural, 23; and nonmonetized value, 130; social, 132; tourism as earning, 75; voluntourism as "sink" for, 20–21, 147
capitalism: affect in, 25, 167–68, 172n14; and coloniality, 20–21; and creative destruction, 147; effect on Utila, 32; impact on environment, 18; and market approaches to conservation, 23–24, 115–16, 164–66; new materialism on, 33; transformation of nature in, 115–18, 127
"Caranguejos com Cérebro" ("Crabs with brains") manifesto, 64–65

carbon trading, 23, 115
care, ethic of, 70–71
care work: cultural capital earned through, 138; death and killing in, 140–41, 157; feminism on, 24–25; in multispecies conservation voluntourism, 33, 127, 171n6; in social hierarchy, 149; in stewardship science, 146; by voluntourists, 117, 125–27, 131
Caribbean, the: English language in, 11–12; feminization of, 75–76; Internet images of, 73–74; lionfish in, 15; neoliberalism in, 95–96; as "paradise," 109–11; and rhizomatic thinking, 63–64; tourism advertising of, 109
Carib Indigenous people, 9–10, 46
Carnival, as tourist binge, 79, 80, 94, 176n7
Cartesian human/nature divide, 21–22, 116–17
Cayan identity and people, 12, 44, 53, 143–44
cetourism, 115–16
"Chasing Whales with Bateson and Daniel" (Neves), 151–52
Chthulucene, the, 168, 170
Clayton-Bulwer Treaty, 12
coconut bug, 3, 170
coconuts: in beach environment, 70, 72, 96–97; cultivation of, 12, 45, 59, 170, 175n13; in food tradition on Utila, 46; in HGTV episode, 32; and insects, 3, 170; oil production from, 13–14, 96
collaborative survival, 168–70
colonialism: discourse on race of, 51; and dispossession of Indigenous people, 9, 46–52; and Internet images, 73–74
coloniality: and movement, 30–31; and naming conventions, 30; and ruination, 33; and voluntourism, 20–21
"colored" ethnic designation, 10
Columbus, Bartholomew, 48
Columbus, Christopher, 47, 48
Columbus, Ferdinand, 47

communitas concept, and tourist experience, 74–75
conch, commercial diving for, 143
conservation: affective labor of, 152–53; command-and-control management in, 164; and cultural values, 28; and development, 17; and dispossession, 20; neoliberal and market approaches to, 23–24, 115–16, 172n12; as restriction, 72, 74, 118–19; and tourism development, 95–100, 111; as tourist imaginary, 146–47
conservation organizations: as "affect generators," 164–65; dependence on voluntourist fees, 17–18, 29, 145–46, 163–64, 173n16; images from, 149; kill campaigns by, 69, 138–41, 179n21; local awareness of, 157–58; marketing of dive tourism, 132–35; as NGOs, 16, 123; as private businesses, 16; reducing local freedoms, 118–19; and regulatory codes, 147; reliance on social media, 163; staff turnover in, 29, 129, 134–35, 145; underfunding of, 17–18, 29, 123–24; on Utila, 16, 29–31, 118–66, 173n19
conservation voluntourism: in affect economy, 116–18; as affective labor, 22–27, 95, 148; as Anthropocene tourism, 113; as capitalist endeavor, 19–20, 147; and coloniality, 20–21; and creative destruction, 147; definition of, 4–5; "Disneyization" in 164–66; and emergent ecologies, 152–53; imaginaries in, 135–36; as industry, 111, 148, 149; lack of evaluation research on, 145–46; motivations for taking part in, 25, 136, 179n18; and neoliberalism, 23–24, 129–32, 173n16; political ecology on, 18–22, 148–66; as service industry, 115; as "sink" for capital, 20–21, 147; as step toward employment, 26, 29, 130–32, 165–66; recruitment in, 16, 145, 163–66; subjectivities created in,
114, 147; suffering subject as generator of, 117, 150; on Utila, *see* conservation voluntourism on Utila. *See also* volunteers; voluntourists
conservation voluntourism on Utila: experiences with local organizations in, 124–27, 136–37, 145, 153–66, 173n19, 179n14, 179n19, 179n21; origins of, 30–33; party scene and, 32, 91–95
consumption: as activism, 164; dedifferentiation of, 165–66; of experiences, 163
Cooper, Alton, 76
Cooper-Key, Captain, 12–13
coral: bleaching of, 118; and mangroves, 64; reduction by lionfish, 15, 139; spearing of, 141, 147, 179n21
coral reef systems: conservation of, 16; and dive buoys, 119–20; and lionfish, 69, 139–41; mesophotic, 139; species inhabiting, 15; Utila and, 4
Coral Beach Village, 17
Coral Reef Alliance, 16
corruption: and housing development, 102; and illegal logging, 17; and local police, 90–91; and political instability, 32
Creole, the, 51, 71–72
"Creole" ethnic designation, 10
Creole language, 11
creolization: in identity, 51; of language, 11
Crusoe, Robinson, character. *See Robinson Crusoe* (Defoe)
cultural values, local, 28

data: collection as affective labor, 137; collection as labor crafted for volunteers, 157, 163–64; intrusive practices in collection of, 7, 147, 152–53, 157–63, 166, 171n7; on lionfish, 139–41; on swampers, 7
Davidson, William, review of historical sources, 49, 50
death and killing: in affect economies, 20, 140–41, 157, 168; as central to con-

Index

servation voluntourism, 157, 146, 153, 157, 179n21; of insects, 157; of lionfish, 69, 138–41, 146, 168; of Pech Indigenous people, 9, 45, 46, 48–49
De Avila, Francisco, 49–50
Debord, Guy, on spectacle, 117
decompression sickness (DCS), 141–46
Deumert, Ana: on landscape assemblages, 22; on "think with the mangrove," 64, 170
development: and claims on nature, 18; clearing of mangroves for, 15; and conservation, 17; dominant models of, 18
discovery, discourses of, 20, 88, 111
"Discovery Channel Utila," 153
"Disneyization," 164–66
dispossession: in affect economy, 132; and gender, 108–9, 177n19; of human and nonhuman species, 169–70; of lobster divers, 141–46; of Pech Indigenous people, 45, 77; rhetorical, 20, 78–79, 142; and tourism development, 100–111; on Utila, 77–79, 111, 148
distinction, earned in tourism, 75
dive shops, 83, 90–91, 98, 149, 152, 173n20
dive tourism: and affect economy, 150; conservation organization marketing of, 132–35; foreign-owned businesses in, 99; marketing of, 76, 79–88; party lifestyle and, 79–88, 90–91; multispecies encounters in, 141; number of people taking part in, 149; privileged over local fishing, 119–20; rise of on Utila, 32–33, 97–98; and undersea life, 76
"Dr. John," as YouTube celebrity, 32, 79–82, 80, 94
DuBois, W. E. B., on white privilege, 32, 109
duppy/duppie/duppa spirits, 40–41, 45
DuPuis, John, 80–81, 176n8

ecosystem services, payment for (PES), 23, 115
ecologies, emergent, 152–53

ecotourism, 115, 178n5
employment: flexible, 131; offshore, 14; in unstable organizations, 29; voluntourism as step toward, 26, 29, 130–32, 165–66
Engels, Jim, 133, 134, 141
English language, 10, 11–12
entanglement: in affect economy, 150–53; in historical fiction, 61–62; in multispecies relationality, 21–22, 24, 33, 52, 116, 133, 149–50, 170; rhizomatic thinking and, 63–64
entrainment, of human and non-human, 151–53, 166
environmental governance, 23
Escobar, Arturo: on plantation form, 50; on political ontology, 21
ethnicity: on Utila, 8–11, 66–69, 172n8; in political ecology, 18–19
ethnography: multispecies, 115–18; on Utila, 8
Eurocentrism: and history, 169; and knowledge, 29–30; and "Man," 51–52, 175n9
Evans, David, 10
Everglades, 69
exchange value, 19–20, 116, 138, 149, 155, 164–66, 168–69, 172n14
Executive Agreement 005–97, 17
Executive Agreement 142–2009, 17
Export Processing Zones, 95

Facebook, 137, 146, 152, 179n23
feminism: "affective turn" and, 167–68; on care work, 24–25; decolonial, 167–68; in political ecology, 29–30
fetishization, of nature as commodity, 115–16
fieldwork: in GCS framework, with local people, 114–15; as methodology, 27–28
fisheries, commercial: for conch, 144; dispossession of, 149; "fish factories" in, 145; for lobster, 14, 141–46; for shrimp, 14; zoned categories for, 17

fishing, traditional/subsistance: affect in, 138; dispossession of, 149; dive tourism privileged over, 119; made illegal, 127; and offshore employment, 14; and "Old Tom" (whale shark), 112, 115, 137–38, 149; zoned categories for, 17

"food to friend" sea turtle message, 38–39

food traditions, Utilian, 8, 17, 37–39, 46, 100, 112, 125–26, 127–28, 147, 149–50, 164, 169, 180n3

Foucault, Michel, and *homo economicus*, 133–31

freedom: "bush" as zone of, 71–72, 176n3; from restrictions, 74; from rules, 79, 163; and tourism, 72–79, 111; and "untouched" marine environments, 73; Utilian view of, 71–72, 175n2

Fundación Islas de la Bahía (FIB): as NGO, 16; management of Iguana Station by, 124; as protected area comanager, 17; staff turnover at, 134–35; terrestrial program, 17–18

fundraising. *See* resources for conservation organizations

gap year, voluntourism in, 26–27, 131

Garifuna Afro-Indigenous people: history of, 9–10, 177n19; individual identifying as, 79, 110; land rights of, 28

gender: and care work, 24–25; in conservation voluntourism, 18, 171n6; and dispossession, 108–9, 177n19; in historical fiction, 56–61; in political ecology, 18–19; and tourism, 75–76; and rape, 89–90

Generation Y, and voluntourism, 26–27

Ginsky (video maker), 84–85

Giumbe (Indigenous translator), 48

Glissant Édouard, on rhizomatic thinking, 63–64

Global Change Science (GCS): Moore on, 113, 177n1; research on local people in, 114–15; Utila as laboratory for, 113, 114, 163

Global North: migrants from in Utila, 123; "otherness" reinforcing dominance of, 25–26; privileged subjectivity of, 132

Global South: impact of tourism development on, 113; in voluntourist social media posts, 117

GoFundMe campaign for lobster divers, 142–46, 179nn23–24

Google, 76

Graburn, Nelson, on imaginaries, 135–36

Graham, Ross, on Creole speakers, 11

green economy: and affective labor, 19, 20, 24, 166; consumption in, 164; global tourism in, 19, 113–14

green iguana (*Iguana iguana*), 15, 128

green sea turtle (*Chelonia mydas*): as endangered, 15; meat of, 150

Guanaja: BICA on, 16; ethnicity on, 10; Spanish and, 47, 49–50; waters around, 17

Hardt, Michael, on immaterial labor, 24

Hairy Mango, as curative fruit, 66

Half Price Paradise Utila Honduras (HGTV), 78–79

Haraway, Donna: on "becoming with," 8, 167; on Chthulucene, 168, 170

hawksbill sea turtle (*Eretmochelys imbricate*): beach nesting sites of, 39; as endangered, 15

hedonism, and tourism, 75, 79

"highlander" iguana (*Ctenosaura similis*): mangrove habitat of, 15; as non-endangered, 15; proliferation of, 128–29; toe removal from, 158–60, *159*

HMS *Pscyhe*, 12–13

Home and Garden Television (HGTV), 32, 78–7, 100, 101–2, 176n6

homo economicus, 130–31

Honduran Congress, 17, 95

Honduran Constitution, Article 107, 105
Honduran people: ethnicities of, 9–11; facilitating research, 147; and "Spaniard" ethnic designation, 68; as student volunteers, 16–17
Honduras, Republic of: affect economy in, 20; Bay Islands as Department of, 12–13; Callejas administration of, 105; coup of 2009, 32, 108, 174n22; living standard in, 110; lobster diver deaths in, 142–43; national marine park, 17; Spanish as official language, 11; Tourism Free Zones in, 16–17, 95–96, 100, 177n15; Utila ceded to, 31
Honduras Travel, 80–81
hotels: vs. boardinghouses, 85; bookings during Semana Santa, 99; foreign ownership of, 98; increased number of, 98–99
humans: affective encounters with other species, 19–20, 150; assemblages with other species, 69; "becoming with" by, 167; Cartesian divide from nature, 21–22, 116–17; creation of Anthropocene era by, 113; entrainment with nonhumans, 151–53, 166; as *homo economicus*, 130–31; as relational with landscape, 52; separation from other forms of life, 146–47
House Hunters International Caribbean Life (HGTV), 78–79, 101–2
housing: and available land, 148–49; construction of, 39, 101–104, 106–109, 135, 173n4; low-cost, 106
hyperbaric chamber: decompression sickness treated by, 141–42; GoFundMe campaign and, 142–46

identity: as British, 12–13, 68; and care for turtles, 38; and color, 10–11; Pech, 77; voluntourist, 26–27
"If you come to Utila, you're never going to leave" slogan, 69

"If you come to Utila, you can do what you want" (MartyandGinski) video, 80–88, *80*, *83*, 95, 110
Igoe, Jim: on environmental impact of tourism, 23, 32; on making nature move, 127; on white privilege, 32
iguanas: eating crabs, 127; hunting of, 125; reduction of habitat, 98; and mangroves, 64, 126, 127–28; and racoons, 168–69, 170; restriction of local activities around, 125–26, 127, 149; as suffering subject, 117; value in survival of, 168–69; voluntourist care for, 125–27. *See also* "swamper" iguana; green iguana; "highlander" iguana
Iguana Research and Breeding Station (Iguana Station): dependence on volunteer fees, 124, 129–30; environmental education programs of, 124; founding of, 125; management by FIB, 124; as NGO, 16; staff turnover at, 130, 179n17; "swamper" projects of, 124; as underfunded, 123, 124, 129; volunteer experience at, 124–27, 130, 173n19
images: on Internet, 32, 73, 79; of multi-species encounters, 150–52; in social media, 27, 94, 157, 163–66, 173n18; as spectacle, 117; theming of, 165–66; of voluntourism, 27, 149, 154, 157
imaginaries in conservation voluntourism, 135–36, 146–47
"Indian Well" site, 69
Indigeneity, colonial ideas about, 61
infilling of coastline, 103, 106
Instagram, 146, 152, 173n18
Inter-American Convention for the Protection and Conservation of Sea Turtles (IAC), 121–22, 174n2
International House Hunters (HGTV), 32
International Union for the Conservation of Nature (IUCN): data funneled to, 164; protected areas and, 17; Red List of endangered species, 15

International Whale Shark Day, 132–33
internships, fees for, 133–35
Internet: AboutUtila.com on, 76–77, 79–88; Caribbean images on, 73–74; fee-based internships marketed on, 134–35; Google hits on, 76; tourism images on, 32; YouTube video of Utila party scene on, 32–33, 80–88, 95, 110
Isabella, Queen, 47

Jackson, William: historical fiction by, 53–63; on Pech Indigenous people, 77
Jericho neighborhood, 44
Jim Engels Ocean Steward Scholarship (JEOS), 134
Johnson, Melissa: on the Creole, 51, 175n9; on freedom, 71–72, 74

Kanahau Conservation Research Facility: bat research of, 40; as NGO, 16; terrestrial program, 17–18
Kent, Suzanne, collaborative research on Utila, 28–29, 91–92, 178n3

labor: crafted for volunteers, 155; of early Utilians, 12–14; emotional, 165–66; enslaved, 109–110; immaterial, 24–27; of "protection," 127; in social hierarchy, 149; value of, 24; volunteer vs. paid local, 25, 129–30, 147
landscape: assemblage in, 22, 45–46, 52; coloniality in, 22; "one-world worldview" shaping of, 42; plantation, 109–10; and rhizomatic thinking, 63–64; shaped by voluntourists, 146
Las Casas, Bartolomé de, 47–48
Latour, Bruno, on affect, 19
law, English, 12
law, Honduran: on beach privatization, 105–6; on protected species and areas, 17, 100–101, 108, 146–47; foreign input on, 147; on foreign ownership, 105–6;

on tourism development, 95–96, 177n15, 177n17
Law of Tourism Development, 95
Law, John, on "one-world world" ontology, 22
"Lawyer Trades City Life for Relaxing Beaches in Utila" (HGTV), 78, 101–2
Le Guin, Ursula, on entrainment, 151–53
"lifestyle migrants," 14, 148
lifeworld: reciprocal exchanges with, 169–70; relationality and, 64; in Utila, 111
liminal state, and tourist experience, 74–75, 163
lionfish: cook-off competitions with, 140, 168; killing of, 138–41, 168; life cycle of, 139; organizations collaborating on, 18, 138; reduction of native species by, 15; regulatory codes on, 147
lobster: commercial fishing of 14, 141–46; protection of juveniles, 17
lobster divers, decompression sickness and, 141–46
loggerhead sea turtle (*Caretta caretta*): beach nesting sites of, 39; as endangered, 1
Lord, David: on coconut oil production, 13–14; on Utilian economy, 66–67
Luxemburg, Rosa, on capitalism, 20
Lyons, Kevin, on neoliberal voluntourism 26–27

MacCannell, Dean, on tourist motivations, 73
mangos: in historical fiction, 53–54, 65; symbolism of, 65–66
mangroves: and blue crab, 62, 127; clearing of for development, 15, 101, *103*; importance as ecosystem, 62–63, 64; four species of, 14–15, 161–62; in historical fiction, 60–62, 175n13; illegal logging of, 17; as prime real estate, 148; propagules of, 162, *162*, 169, 180n2; as refuge, 63, 64; relational ontology of,

36; and rhizomatic thinking, 63–64. *See also* black mangrove; buttonwood mangrove; red mangrove; "think with the mangrove"; white mangrove,

marketing: affective encounters in, 150–51; Caribbean in, 75–76; of experiences, 163–64; for fee-based internships, 134–35; for gap year programs, 131; "otherness" in, 25–26; party lifestyle in, 79–88; tracking of, 28

Marty (filmmaker), 82–83

Marx, Karl: on fetishization, 115; on value, 172n14

matsutake mushrooms, harvesting of, 167–68

Maturana, Humberto, on "biology of love," 30, 169

McMansions, 110

McNab, Shelby: as BICA director, 118–19, 121–22, 179n12; and Brandon Hill Cave, 41; on cannibalism, 46; on duppie spirits, 40, 64; on Pech Indigenous people, 77; on Robinson Crusoe, 34–39, 40, 41, 64, 118, 175n7

Mesoamerican Barrier Reef, 4, 15

mesophotic ecosystems, 139

mestiza/o people: Honduran, 9, 68, 172n8; and "Spaniard" ethnic designation, 10, 52, 68

methodology: feminist, 29–30; fieldwork and, 27–28; political ecology in, 29–31

migrant workers, 9–10

monitoring of species: limited funding for, 17; of orchid bees, 6; of seagrass, 146; of swampers, 147, 157–63; of turtles, 18, 38, 118, 122, 152, 156, 174n4; of whale sharks, 115, 134

Monroe Doctrine, 12

Moore, Amelia: on Anthropocene era, 113, 123; on Global Change Science, 113, 177n1

"more-than-human," the: and affect, 19; in Anthropocene, 20–22; as world, 8

Mostafanezhad, Mary, 18–19

movement: and coloniality, 30–31; of nature, 127; of species, 33

movimento mangue (mangrove movement) cultural-political movement, 64–65

multispecies encounters: affect in, 19–20; and anthropology, 114–18; and assemblage, 8, 22, 52, 64, 69, 111, 138, 149–53, 169–70; "becoming with" and, 116–17, 148; and capital, 149; collaborative survival in, 168–70; as entanglement, 21–22, 24, 33, 52, 63–64, 116, 133, 149–50, 170; entrainment in, 151–53, 166; ethnography on, 115–18; exchange value of, 165–66; inequality in, 33, 146–47; islander stories of, 28; as kin, 149; purchasing of, 165–66; rhizomatic thinking and, 63–64; as spectacle, 150–53

Mushroom at the End of the World, The (Tsing), 167–68

Nanci (Utila resident), 106–9, 145

Naomi (Utila resident), 96–97

nature: "bush" as, 71–72; Cartesian divide from human, 21–22, 116–17; claims to, 18; as commodity, 115–16, 127; desire for, 165–66; market approaches to, 23–24, 115, 131–32, 164; "productive use of," 169–70; transformation under capitalism, 115–18

natures, new, 18, 24–25

Nature™ *Inc.* (Fletcher, Dressler, and Büsher, eds.), 23–24

neoliberalism: in Caribbean, 95–96; effect on Utila, 32; and environmental governance, 23–24, 172n12; as subjectivity, 123; and tourism development, 95–100; voluntourism as, 26–27, 129–32

Neves, Katja: on cetourism, 115–16; on whale/whaler entrainment, 151–52

new materialism: and "biology of love," 169; and colonial capitalist ruination, 33; and "more-than-human," 20; in political ecology, 20–22
New Orleans, as "little Utila," 12
nostalgia, 14, 176n13

Ogden, Laura: on landscape assemblages, 22; on mangrove and rhizomatic thinking, 63, 69
"Old Heads" (settler families): and boas, 153–55; and dispossession, 103–4, 108–9; and iguanas, 128–29; honoring of Pech Indigenous people, 77; language of, 11–12; and McNab pamphlet, 35; oral traditions of, 40–42, 64; and Rose book, 29
"Old Tom" (whale shark). *See* whale shark ("Old Tom")
Olimpia massacre: events of, 53; in historical fiction, 54–60, 65; islander accounts of , 66–67
one-sided communication, 150–52
"one-world world" ontology, 22, 50, 64, 148, 169
ontology: "more than human," 21; of "one-world world," 50, 64, 148, 169; political, 21–22, 30, 45–46, 64, 169–70; relational, 36
Operation Wallacea (Opwall): in lionfish collaboration, 18, 138; lionfish ecology program of, 134; as private business, 16; spearfishing forbidden to volunteers by, 138, 141, 179n21
orchids, 3, 5
orchid bee (*Euglossa*): death of, 6, 146, 170; trapping of, 5–6, 7, 156, 161, 167
Original Volunteers holiday, 124
Oyster Bed Lagoon project, 17

"Paya". *See* Pech Indigenous people
Päyä: The Roatan Lifestyle Magazine, 77
"Paya Resistence" (Thompson), 77

payments for ecological services (PES), 115
Pech Indigenous people: archeological record and, 43–46; in Bay Islands, 9; decimation of, 9, 45, 46, 111; dispossession of, 45–52; enslavement of, 32, 46–47, 48; landscape assemblages and, 46; multispecies interactions of, 169; tales about, 41, 45, 46–47; and "think with the mangrove," 64; trade with Yucatán, 47–48
Pederson, Glenn, 122–23
pirates: in Bay Islands, 9, 49–50; tales about, 41; on Utila, 31
plantation: banana, 9–10; coconut, 45; economy as, 111; Escobar on, 50; as landscape, 109–10; migrant workers and, 9–10; and tourism, 73–74, 109–10
Plantation Beach Resort, 74
political ecology: on affect, 31–32, 166; core elements of, 18–19; feminist, 29–30; of multispecies conservation voluntourism, 18–22, 148–66; new materialism in, 20–22; political ontology in, 21–22; of tourism, 18, 114
political ontology: feminist, 30; and historical fiction, 64; Indigenous dispassion viewed through, 45–46; power in, 21–22, 169–70
privatization, 71, 104
Professional Association of Diving Instructors (PADI), 82–83, 143
protected areas: contested boundaries of, 33, 146–47; and IUCN classifications, 17; laws on, 100–10; lionfish culling and, 138–41; management of, 8, 17, 138; nature as commodity in, 127; resident attitudes on, 99–100, 127–29; voluntourists permitted into, 163
"protection," discourse of, 33, 69, 127, 146–47, 149
ProTECTOR, 174n4
Pumpkin Hill: blue crabs at, 76; Robinson Crusoe and, 36–37, 39

Pumpkin Hill Beach: development of, 148–49, 173n4; freshwater cave near, 155–56

race: colonialist discourse on, 51–52; Enlightenment ordering of, 52; in historical fiction, 54–63; relations in, 52–54, 63–66; social stratification through, 52, 66–69

racoons: and iguanas, 160; introduction of to Utila, 128; survival with other species, 168–69, 170

rape, 89–90

red mangrove (*Rhizophora mangle*): propagules of, 180n2; as swamper iguana habitat, 14–15; salt tolerance of, 161–62. *See also* mangroves

reef ecosystem. *See* coral reef systems

Reef Fee, 141, 173n20, 179n22

Reef Leader program, 122, 128, 178n11

regulatory codes, written with foreign researchers, 147

Rehab (bar), 79

relationality: and moving beyond dualisms, 21–22; one-sided communication in, 150; as reshaping space, 52; and rhizomatic thinking, 63–64; synchronization in, 151–53; worlding as framework for, 149–50

remittance economy, 66

resources for conservation organizations: grants as, 123, 164, 173n16; need driving data collection, 163–64; soft money as, 29; and underfunding of local organizations, 17–18, 123, 124, 129; voluntourists as source of, 17–18, 29, 123, 129–30, 145–46, 163–64, 173n16

rhizome, Deleuzian, 63

rhizomatic thinking, 63–64

rites of passage, 74–75, 177n2

Roatán: BICA on, 16–17; ethnicity on, 10; in historical fiction, 53; pirates on, 50; plantation resort on, 74; waters around, 17

Robinson Crusoe (Defoe): exhibit on, 35, 37; Utila as purported setting for, 34–39, 40, 42, 46, 118, 175n7

Rock Harbour and Raggedy Cay marine area, 17

Rose, Richard H.: on Brandon Hill Cave, 41; descriptions of Utila, 3–4, 30, 45–46, 51–52, 55; on "Indian Well" site, 69; on mound excavations, 43; and "Old Heads," 29; on Pech Indigenous people, 77

sacred journey, tourism as, 75

Salazar, Noel, on imaginaries, 135–36

Sandy Bay West End Marine Reserve, 174n4

"Saving Lobster Divers" GoFundMe campaign, 141–46, 179nn23–24

savior discourse, 142

Sayda (Utila resident), 62–63

Schumpeter, Joseph, on creative destruction, 147

scuba certification and training, 4, 28, 82–83, 133, 134, 143–44

scuba diving: and coral reef systems, 119–20, 141–42; foreign expatriates in industry, 14; by lobster divers, 143–46

Seagrape Plantation Resort, 74

Sea Turtle Conservancy, 174n4

sea turtles: endangered species of, 15; exhumation and release of, 122, *122*; "food to friend" campaign for, 38; in historical fiction, 65; historical accounts of, 36–37; hunting of, 17, 149–50; monitoring of, 18, 38, 118, 122, 152, 156, 174n4; nesting of, 15, 39, 69, 155; restriction of local activities around, 17, 120–23, 125; as suffering subject, 117; Utilian raising of young, 37–38, 121–22, 150; value in survival of, 168–69. *See also* green sea turtle; hawksbill sea turtle; loggerhead sea turtle

Second World War, 14
Semana Santa holiday: eating of iguana eggs at, 161; hotel bookings during, 99
sentient beings, non-human, 150
settler ideology: and segregation, 10; and worlding, 111
sharks: in historical fiction, 59, 65; whale sharks as, 132
Singh, Neera, on affective labor, 147
Skid Row (bar), 84
slavery: in historical fiction, 66; immigration of freed people, 9; and Indigenous people, 32, 46–47, 48, 77; and tourism, 109–10
Smith, A. E. B., 40–41, 53
Smithsonian Institution archeological study, 42, 45
socialities: and affective labor, 147; and multispecies interaction, 20; new, 69; "nonserious," 75
social media: "biography" in, 27; care work in, 113; exchange value of, 165–66; fee-based internships in, 133–34, *134*; fundraising through, 142–46; lionfish culling in, 140–41; as one-sided communication, 150–52; recruitment of conservation voluntourists through, 145–46; savior discourse in, 141; tourism imaginaries in, 135–36; voluntourism images in, 27, 94, 157, *157*, 163–66, 173n18; water as favored setting in, 163; wildlife encounter videos in, 137
socionatures: and conservation voluntourism, 24–25; precolonial, 51–52
Southwest Cay marine area, 17
Spain: "Carib" designation by, 46–47; Christianity of, 47; conflict with British, 9, 31, 34, 49–50, 52–53; decimation of Pech under, 9, 45, 46–49; introduction of mangoes, 65
"Spaniard" ethnic designation, 10, 52, 68, 143

Spanish language, 10, 11–12
spearfishing: injury to coral by, 141, 147, 179n21; for lionfish, 147, 179n21; national park ban on, 17; training and permits for, 138
species: affect among, 19–20; entrainment among, 151–53; inequalities among, 33, 146–47; life and death choices between, 19–20, 20, 33, 140, 157, 170; movement of, 33; as suffering subjects 117–18, 150
species banking, 23, 115
spectacle: affective encounters with other species in, 19–20, 148, 150–53; conditioning by, 166; image concealing inequities as, 117; lionfish management as, 140–41
sperm whales, 151–52
Standard Fruit Company, 12
status, earned in tourism, 75
Staying with Trouble (Haraway), 168
stewardship science, 146
Stonich, Susan: on BICA, 16–17; on Spanish speakers, 11
Strong, William Duncan, 42, 45
subjectivities: affective labor transforming, 14, 147; neoliberal, 123, 130–32; of suffering, 117, 150; tourism imaginaries in, 135–36; voluntourist, 26–27, 114, 124
Sullivan, Sian: on environmental impact of tourism, 23; on making nature move, 127
Sundberg, Juanita, on Eurocentric knowledge, 29–30
"Sunjam" rave, 79, 87–88
surf tourists, motivations of, 73
"swamper" iguana (*Ctenosaura bakeri*): breeding programs for, 164; colors of, 160–61; as commodity for "protection," 127; as endangered, 15, 69; hunting and egg harvesting of banned, 17, 125, 147, 161, 164; measurement methods for, 160–61; reduction of mangrove

habitat, 15, 69, 103, 126, 169; toe removal for data collection, 7, 147, 158; tracking of, 160, 164, 171n7, 173n19; volunteer projects focused on, 17–18, 126, 157–63
Swamplife (Ogden), 69
Swan's Bay, airport site at, 98
sweep netting, 156–57

tarantulas, in residences, 153
"Telling Is Listening" (*Le Guin*), 151–53
"think with the mangrove": and historical fiction, 65; and rhizomatic thinking, 63–64; Utilians and, 65–66
"third world" label, 88
ticks: on boas, 153, *154*; as disease vectors, 154–55, 180n1
"TNT" (AC/DC), 81
tourism: and affect economy, 69; and freedom, 72–79; and hedonism, 75, 79; imaginaries in, 135–36, 146–47; Internet images of, 32, 72; literature on, 72–75; motivations for, 72–79; political ecology of, 18, 114; as rite of passage, 74–75, 177n2
tourism development: and available land, 148–49; and conservation, 95–100; and dispossession, 100–11; in Global South, 113; and "protection" discourse, 69; and social hierarchy, 149
Tourism Free Zones in Honduras: benefit to investors, 95–96; displacement by, 23; creation by Honduras, 16–17, 95, 177n15; and tourist development, 100
tourism on Utila: boom in 1980s, 14; crime and, 89–91; development due to, 95–111; and dispossession, 77–79, 100–11; economic dependence on, 97; party lifestyle and, 79–88; resident attitudes on, 71, 85–91, 95–100; "rules do not apply" image and, 75–76, 79; season for, 8; swimming with whale sharks, 112; as unregulated, 33

tracking: of swampers, 160, 164, 171n7, 173n19; of whale shark, 18, 115
Tradewinds housing development, 101–4, *102*, 110
Tranquila Bar, 83
tributarios (tribute-paying Indigenous people), 49
Tsing, Anna, 167–68
tuc tuc (motorized rickshaw), 89
Turner, Victor and Edith, on rites of passage, 74–75
Turtle Harbour marine area, 17, 178n10
Turtle Harbour terrestrial wildlife refuge, 17, 157–58, 160
turtle monitoring, 18, 38, 118, 122, 152, 156, 174n4
turtling tradition in Utila, 37–38, 121–22, 150
Twitter, 146, 164

Unidad Municipal Ambiental (UMA): founding of, 16; as protected area comanager, 17
United Fruit Company, 12
United States: consulate of, 91, 176n12; fruit companies of, 9, 12, 13; and Monroe Doctrine, 12; race in, 55; shipping from, 12; standard of living in, 110
"untouched"/"undiscovered" labels on environments, 20, 73, 88
Utila: affect economy of, 37–38, 69, 138, 140, 163, 167–70; American consulate on, 91; archeology of, 42–46; conservation voluntourism on, *see* conservation voluntourism on Utila; drug use on, 87, 89–91; early economy of, 12–14; electricity on, 4, 36, 99, 85, 101, 149; environmental concerns of, 15–16, future of, 167–70; as Global Change Science laboratory, 113, 114–15, 163; protected area boundaries, 33; residential segregation in, 10–11, 54, 63; as Robinson Crusoe's Island,

Utila (*continued*)
34–39, 40, 42, 46, 118, 175n7; Rose on, 3–4; tourism in, *see* tourism on Utila; transportation on, 99; waters around, 16–17, 118, 173n20

Utila Hyperbaric Chamber, 142–43

Utila Lodge, 133, 142

Utila: Past and Present (Rose): on environment, 3–4, 12–13, 29–30, 32, 41, 43–44, 46, 51; and "Old Heads," 29

Utilian people: attitudes on tourism, 71, 85–91, 95–100; awareness of conservation organizations, 157–58; and Bay Islands English, 11–12; dispossession of, 77–79, 111, 148; and ethic of care, 70–71; ethnicities of, 9–11, 66–69; on freedom, 71–72, 175n2; on island's history, 32, 42–46, 169–70; knowledge of local species, 37–38, 121–22, 126–31, 150, 160–61, 178n3; as lobster divers, 142–46; and mangroves, 65–66, 69; protected species in food traditions of, 8, 17, 37–39, 100, 112, 147, 149–50, 161, 164, 169, 180n3; subjectivities of, 112; voluntourist views of, 126–28

value: of affective labor, 127, 164–65, 168; as exchange, 19–20, 116, 138, 149, 155, 164–66, 168–69, 172n14; of experiences, 146; nonmonetized, 130; produced through interactions with nature, 24, 115–18, 139–40; in survival of species, 168–69

van Gennep, Arnold, on rites of passage, 74–75

Verden-Zöller, Gerda, on "biology of love," 30, 169

"Visit to Dr. John in Utila, A" (DuPuis), 80–81

volunteers: experience of, 5–8; motivations of, 25–26; recruitment of, 16, 145; fees from as revenue source, 17–18, 29, 173n16; local people as, 122.

See also conservation voluntourism; voluntourists

voluntourists: and affective encounters, 150–51; consumption of digital sources, 152; crafting of experiences for, 155–57, 163; experiences with local organizations, 124–27, 136–37, 145, 153–66, 173n19, 179n14, 179n19, 179n21; identity of, 26–27; lack of engagement with local community, 145–46; vs. local labor, 129–30; motivations of, 25–26, 29, 130–32, 136, 165–66, 179n18; recruitment of, 16, 145, 163–66; as neoliberal citizens, 129–32; as revenue source for organizations, 17–18, 29, 123, 129–30, 145–46, 163–64, 173n16; spearfishing forbidden to, 138, 141, 179n21; subjectivity of, 124; and suffering subject, 117, 150; views on Utilians, 126–28. *See also* conservation voluntourism; volunteers

Vrasti, Wanda, on neoliberal subject, 130–32

water quality management, 118

Wearing, Stephen, on volunteer tourism, 4–5

West, Paige: on capitalism, 20; on tourist motivations, 73

whale shark ("Old Tom") (*Rhincodon typus*): bonito tuna drawn by activity of, 112, 115, 133, 136–37; conservation organizations facilitating encounters with, 19, 115–16, 133–40, 147; as endangered, 15; regulatory codes on, 147; stress from tourist proximity, 136; as suffering subject, 117; as tourist attraction, 112, 115, 132–40; tracking of, 18, 115; year-round population at Utila, 15

Whale Shark Oceanic Research Center (WSORC): fee-based internships, 133–

35, 166; fundraising by, 132–33; JEOS scholarship offered by, 134; in lionfish collaboration, 18, 134, 138; mangrove nursery of, 134; as private business, 16, 132; staff turnover at, 134–35; tourist packages offered, 133–35; voluntourist experience with, 135–37, 179n19, 179n21; whale shark tracking by, 18, 134

whale watching: vs. cetourism, 115–16; number of tourists taking part in, 149

"white" ethnic designation: individuals identifying as, 40, 70; Lord on, 66–68; origin of, 67; self-identification as, 10

Whitefield, Dillard, 43–44

white mangrove (*Laguncularia racemosa*), 14–15, 161. *See also* mangroves

Wildbook database, 132, 137

"wishiwilly". *See* "swamper" iguana

worlding: alternative, 166; colonialist, 47; as framework of relationality, 149–50; and modernity, 46; and multispecies entanglement, 33; in political ontology, 21–22; separating humans from other forms of life, 146–47; tourist-centric, 111; Utilian, 72, 111, 169–70

world making: creativity in, 169–70; entanglements in, 31; power in, 21

Wynter, Sylvia, on "Man," 51–52, 175n9

YouTube video on Utila party scene, 32–33, 80–88, 95, 110

Yucatán, Pech trade with, 47–48

About the Author

Keri Vacanti Brondo received her PhD from Michigan State University and is a professor and chair in the Department Anthropology at the University of Memphis. She is a National Geographic Explorer with research interests in feminist political ecology, conservation and development, tourism and local livelihoods, and nature-based volunteerism. She is the author of *Land Grab: Green Neoliberalism, Gender, and Garifuna Resistance in Honduras* (UAP, 2013), and editor of *Cultural Anthropology: Contemporary, Public, and Critical Readings* (Oxford University Press, 2017, 2020), and *Anthropological Theory for the Twenty-First Century: A Critical Approach* (University of Toronto Press, 2021, coedited by A. Lynn Bolles, Ruth Gomberg-Muñoz, and Bernard C. Perley). She has served on the Executive Board of the American Anthropological Association (AAA), as chair of the AAA's Committee on the Status of Women in Anthropology (COSWA), as chair of the AAA's Committee on Practicing, Applied and Public Interest Anthropology (CoPAPIA), as co-chair of the AAA's Members' Programmatic Advisory and Advocacy Committee (M-PAAC), as senior board member of the AAA's Anthropology & Environment Society, as co-chair of the Consortium of Practicing and Applied Anthropology Programs (COPAA), and on the Board of Directors for the Society for Applied Anthropology (SfAA). She is the recipient of the Presidential Award from the American Anthropological Association, the Sierra Club's Dick Mochow Environmental Justice Award, and the University of Memphis Dunavant Faculty Professorship.